T0192425

Tensor Calculus and Applications

Mathematics and Its Applications: Modelling, Engineering, and Social Sciences

Series Editor:
Hemen Dutta

Discrete Mathematical Structures: A Succinct Foundation
Beri Venkatachalapathy Senthil Kumar and Hemen Dutta

Concise Introduction to Logic and Set Theory
Iqbal H. Jebril and Hemen Dutta

Tensor Calculus and Applications: Simplified Tools and Techniques
Bhaben Chandra Kalita

For more information on this series, please visit: www.crcpress.com/Mathematics-and-its-applications/book-series/MES

Tensor Calculus and Applications

Simplified Tools and Techniques

Bhaben Chandra Kalita

CRC Press is an imprint of the
Taylor & Francis Group, an **informa** business

CRC Press
Taylor & Francis Group
6000 Broken Sound Parkway NW, Suite 300
Boca Raton, FL 33487-2742

First issued in paperback 2020

ISBN-13: 978-0-367-13806-6 (hbk)
ISBN-13: 978-0-367-78014-2 (pbk)

Library of Congress Cataloging-in-Publication Data

Names: Kalita, Bharat Chandra, 1937- author.
Title: Tensor calculus and applications : simplified tools and techniques / authored by Bhaben Chandra Kalita.
Description: Boca Raton : Taylor & Francis, [2019] | Includes bibliographical references.
Identifiers: LCCN 2018052527 | ISBN 9780367138066 (hardback :acid-free paper) | ISBN 9780429028670 (e-book)
Subjects: LCSH: Calculus of tensors. | Geometry, Differential.
Classification: LCC QA433 .K345 2019 | DDC 515/.63—dc23
LC record available at https://lccn.loc.gov/2018052527

Visit the Taylor & Francis Web site at
http://www.taylorandfrancis.com

and the CRC Press Web site at
http://www.crcpress.com

Contents

Preface

There is a great demand from students for a book on tensors with simple and conceivable presentations. The theoretical development of the subject "tensor calculus" is critical to be understood by students because of its complex nature of uses of the subscripts and superscripts. The simplification in the working process with repeated/nonrepeated indices of a mixed tensor makes it rather more complex for readers if some special clues are not mentioned. Moreover, the fields of application, namely, non-isotropic media and the exact situation of deformation of bodies, cannot be identified easily in the true sense. Through concrete citation of applicable media and physical bodies, the subject can be made conceivable. For example, in elastic media such as motion of viscous fluids or dirty water, the application of tensors is inevitable. Research on viscous media use of Navier–Stokes equation governed by tensors is one of the primary prerequisites. The investigation causing deformation with elasticity in physics needs application of tensors. However, a clear concept to make use of the different classes of tensors in such fields is of paramount importance. Only then correct results of investigation can be unearthed. Eventually, calculus of tensors can be considered as the most appropriate tool to know the physical field theories. Hence, applied mathematicians, physicists, engineering scientists, and geologists cannot excel without the knowledge of tensors. Emphasis is given primarily on the subject, and only to motivate the readers, some physical fields are described for real interest.

During the past 37 years of teaching this subject at the MSc level, I could clearly read the minds of students on why they found tensor calculus difficult to understand. For the skillful teaching arts adopted in the class lectures, hundreds of students requested that I write a book on tensors for the benefit of students. Students' feedback and suggestions from many colleagues inspired me to undertake this venture of writing this book. With the teaching arts based on some individual special techniques of changing the indices of tensors, I attempted to place a book on tensors in the hands of students. The complexity that arises out of the use of shorthand notations was removed so that readers can easily understand them. If the students capitalize the techniques provided in the book, in addition to its elegant presentation and simple language, my belief will become a reality.

I will be highly obliged if this book renders at least some service to students and researchers so that they can understand its tremendous importance in research gates such as relativity, physics, continuum mechanics, and geology. Overall, this book is designed to cater to the needs of students of mathematics, physics, engineering, and geology from all universities. Further, I will acknowledge the readers with sincere gratitude if they point out mistakes in the formative stage of the book on tensors, which is very difficult to publish without any printing mistakes.

Dr. Bhaben Chandra Kalita
Professor Emeritus
Department of Mathematics
Gauhati University
Guwahati-781014
India

About the Book

The book *Tensor Calculus and Applications* is not elementary in nature; rather, it is physically motivated in the sense of application. Theoretically, the subject "tensor calculus" is critical for students to understand the complex nature of using subscripts and superscripts. Besides the lack of knowledge about the fields of application in non-isotropic media and of identifying deformation situations of bodies poses rather more difficulty to earn the concepts. The elegant nature of description of the theory with specific style of changing suffixes and prefixes and reasons to recover meaningful results of the subsequent fields can only make the subject easier. With this objective in mind, the author was inspired to write the book for the benefit of readers. In the opinion of the author, the old books written by L. P. Eisenhart and C. E. Weatherburn could not serve this purpose though they are of fundamental nature from a theoretical standpoint but not available in the market and conceivable at the same time. The experience derived from teaching the subject for more than 37 years to the postgraduate students and the psychology gathered from the feedback of students are the intense feeling of the author to write a book on tensors using special techniques.

The techniques adopted in the book with directions will definitely encourage the students to read the book to develop concepts and make use of them in appropriate geometrical fields and space. For example, the curvature of space (a geometrical entity) is the manifestation of gravity, and hence, tensor calculus becomes the fundamental tool as discovered by Einstein for general theory of relativity. The book is designed to discuss the fundamental ingredients such as Riemannian tensors, which are essential to enter into the threshold of research in general theory of relativity. Tensor being an intrinsic concept independent of any referential systems, different from Newtonian mechanics, is the essence of invariance for physical laws. Besides, in non-isotropic media such as viscous fluids, elastic media, deformation of bodies similar to structural geology, uses of tensors are essential ingredients. To amplify the uses of tensors in these fields, some relevant ideas are included in the book. In this context, the preface is written to manifest its suitability and necessary background why the author has written the book.

The book consists of 10 chapters. Chapter 1 is devoted to giving some prerequisites of the subjects. Chapter 2 deals with the fundamental concepts of quadratic forms and their properties. Chapter 3 discusses the essential concept of generating space of any dimensions and corresponding geometry like the Riemannian metric inherent in fundamental tensors. Chapter 4 is devoted to developing the subject with the use of shorthand symbols called Christoffel symbols and the important tensorial operation covariant differentiations theoretically. Chapter 5 includes the geometrical concept "geodesics" primarily required for dynamical scenario. Chapter 6 discusses the curvature tensors or Riemannian tensors, the fundamental ingredients of general theory of relativity along with properties. Chapters 7–10 narrate the applications of tensors in general theory of relativity, continuum mechanics, geology, and fluid dynamics, respectively for students.

Author

Dr. Bhaben Chandra Kalita has been first class throughout his career. He has served 37 years in the Department of Mathematics, Gauhati University, in capacity of assistant and associate professors and professor and head of the department since 1978. Dr. Kalita was granted the prestigious award "Professor Emeritus" by the University Grants Commission, Government of India on September 2015. He has published more than 50 papers in *Physics of Fluids, Physics of Plasmas, Astrophysics and Space Science, Journal of Plasma Physics, Physical Society of Japan, Canadian Journal of Physics, IEEE Transaction on Plasma science, Communication in Theoretical Physics,* and *Plasma Physics Reports,* besides some other papers of relativity and graph theory. He has acted as an invited speaker on astrophysics and particle physics in Dallas (2016) and San Antonio (2017). He has presented papers in Granada (Spain), Pissa (Italy), and Swansea (UK), and acted as a speaker in many local universities and institutions. He has also authored several textbooks on advanced mathematics and reference books of higher secondary level. Recently, he has served as 'Keynote Speaker' in *Astrophysics and Particle Physics* conference (2018) held in Chicago, Illinois, USA.

Part I

Formalism of Tensor Calculus

1

Prerequisites for Tensors

1.1 Ideas of Coordinate Systems

Geometric ideas and entities can be well defined in various forms with reference to coordinate systems. In most of the cases for convenience, rectangular coordinate system is taken into account, but it is not applicable in all fields of physical system. Deviating from it, we may think of curvilinear coordinate system in the lowest level. To consider the geometry of space for dynamical scenario as "dynamics deals with the geometry of motion," coordinate system should be suitably selected. For an essential entity in this sense, the author is tempted to give a brief idea of curvilinear coordinate system.

1.2 Curvilinear Coordinates and Contravariant and Covariant Components of a Vector (the Entity)

Consider the rectangular Cartesian coordinates (x, y, z) of any point P with position vector $\vec{r}$. For the coordinates (x, y, z) of P, a correspondence can be made with (u_1, u_2, u_3) as

$$x = x(u_1, u_2, u_3),\ y = y(u_1, u_2, u_3),\ z = z(u_1, u_2, u_3).$$

If these functions are single valued and have continuous partial derivatives, they can be solved as

$$u_1 = u_1\left(x, y, z\right),\ u_2 = u_2\left(x, y, z\right),\ u_3 = u_3\left(x, y, z\right).$$

Here, u_1, u_2, u_3 are called the **curvilinear coordinates** of the point P. Consequently, the position vector $\bar{r} = ix + jy + kz$ can be expressed as $\bar{r} = \bar{r}\ (u_1, u_2, u_3)$.

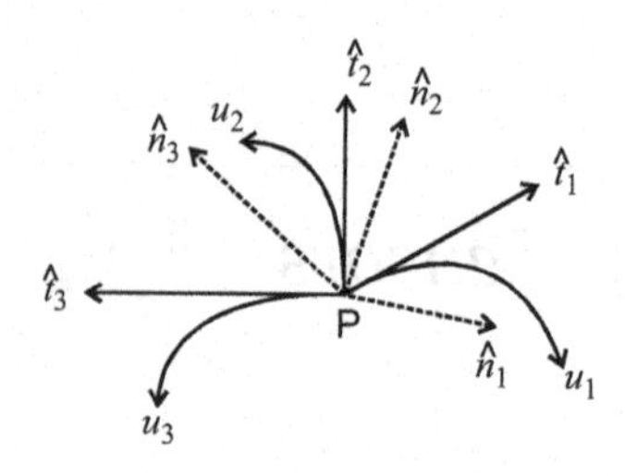

FIGURE 1.1
Vector reprersentations in Curvilinear system.

∴ A unit tangent vector $\hat{t}_1$ to the curve u_1 (i.e., the curve of intersection of $u_2 = c_2$, $u_3 = c_3$) at P (Figure 1.1) is

$$\hat{t}_1 = \frac{\frac{\partial \bar{r}}{\partial u_1}}{\left|\frac{\partial \bar{r}}{\partial u_1}\right|}.$$

Similarly, the unit tangent vectors $\hat{t}_2$ and $\hat{t}_3$ along u_2 and u_3 curves, respectively, can be written as

$$\hat{t}_2 = \frac{\frac{\partial \bar{r}}{\partial u_2}}{\left|\frac{\partial \bar{r}}{\partial u_2}\right|} \text{ and } \hat{t}_3 = \frac{\frac{\partial \bar{r}}{\partial u_3}}{\left|\frac{\partial \bar{r}}{\partial u_3}\right|}$$

Again the normal vectors to the surfaces $u_1 = c_1, u_2 = c_2, u_3 = c_3$ are given by the vectors $\nabla u_1, \nabla u_2, \nabla u_3$, respectively. Hence, the unit normal vectors $\hat{n}_1, \hat{n}_2, \hat{n}_3$ to the direction of the vectors are

$$\hat{n}_1 = \frac{\nabla u_1}{|\nabla u_1|}, \quad \hat{n}_2 = \frac{\nabla u_2}{|\nabla u_2|}, \quad \hat{n}_3 = \frac{\nabla u_3}{|\nabla u_3|}.$$

Thus, at each point P of a curvilinear coordinate system, there exist **two sets of unit vectors**: (i) $\hat{t}_1, \hat{t}_2, \hat{t}_3$ tangent to the coordinate curves and (ii) $\hat{n}_1, \hat{n}_2, \hat{n}_3$ normal to the coordinate surfaces. Of course, the two sets become identical if and only if the curvilinear coordinate system is orthogonal. In this case, the sets are similar to i, j, k of rectangular coordinate system but differ in the sense of changing directions from point to point.

Eventually, any vector $\vec{A}$ can be expressed in terms of the base vectors $\hat{t}_i$ and $\hat{n}_j$ $(i, j = 1, 2, 3)$ as

$$\begin{aligned}\vec{A} &= A_1\hat{t}_1 + A_2\hat{t}_2 + A_3\hat{t}_3 \\ &= a_1\vec{\alpha}_1 + a_2\vec{\alpha}_2 + a_3\vec{\alpha}_3, \quad \text{where } \vec{\alpha}_j = \frac{\partial \bar{r}}{\partial u_j}.\end{aligned} \tag{1.2.1}$$

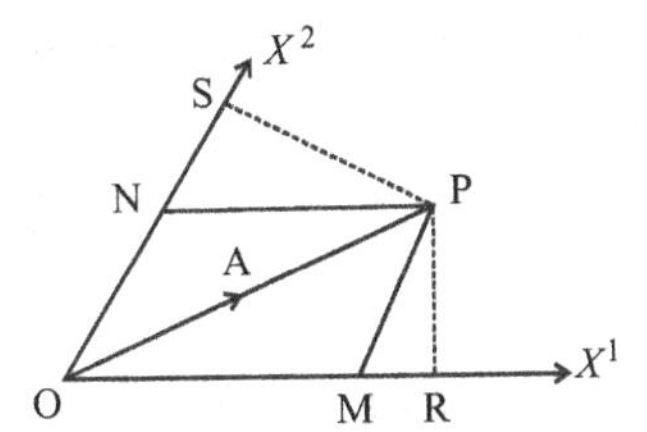

FIGURE 1.2
Vector representations in oblique Cartesian system.

These are not necessarily unit tangent base vectors.

Also,

$$\begin{aligned}\vec{A} &= B_1\hat{n}_1 + B_2\hat{n}_2 + B_3\hat{n}_3 \\ &= b_1\vec{\beta}_1 + b_2\vec{\beta}_2 + b_3\vec{\beta}_3, \quad \text{where} \quad \vec{\beta}_i = \nabla u_i.\end{aligned} \tag{1.2.2}$$

These are not unit normal vectors.

The quantities a_1, a_2, a_3 are called the **contravariant** components, and b_1, b_2, b_3 are called the **covariant** components of the **same** vector $\vec{A}$.

Thus, **based on the representative "base vectors,"** a vector can have contravariant or covariant components.

To illustrate this, let us consider the oblique Cartesian coordinate lines X^1 and X^2 (not rectangular) in two dimensions in a plane.

Let us consider the components OM and ON of any vector $\vec{A} = \overrightarrow{OP}$ measured parallel to the coordinate lines OX^1 and OX^2. They are called the contravariant components of the vector $\vec{A}$ (Figure 1.2).

Consider the perpendicular projections PR on OX^1 and PS on OX^2. OR and OS are called the covariant components of the same vector $\vec{A}$. Obviously, if the coordinate lines are perpendicular, then there will not be any distinction between the contravariant and covariant components.

Depending upon the parallel and perpendicular projections from the point P upon the coordinate axes, we assume the coordinates of P as (x^1, x^2) and (x_1, x_2), respectively.

The contravariant components x^i $(i = 1, 2)$ of the vector $\bar{A}$ and its covariant components x_i $(i = 1, 2)$ are connected by the relations

$$x_1 = x^1 + x^2 \cos\alpha$$

$$x_2 = x^2 + x^1 \cos\alpha$$

where α is the inclination between the coordinate axes OX^1 and OX^2.

$$\therefore \begin{pmatrix} x_1 \\ x_2 \end{pmatrix} = \begin{pmatrix} 1 & \cos\alpha \\ \cos\alpha & 1 \end{pmatrix} \begin{pmatrix} x^1 \\ x^2 \end{pmatrix} \tag{1.2.3}$$

If $\vec{e}_i$ and $\vec{e}_j$ are the unit vectors stipulated along the coordinate axes, then $\bar{A} = x^{1-}\vec{e}_1 + x^{2-}\vec{e}_2 = x^i\vec{e}_i$ so that the length $\overline{OP}$ is

$$|\bar{A}|^2 = \bar{A}\cdot\bar{A} = \left(x^{1-}\vec{e}_1 + x^{2-}\vec{e}_2\right)^2$$

$$= \left(x^1\right)^2 \vec{e}_1\cdot\vec{e}_1 + 2x^1x^2\vec{e}_1\cdot\vec{e}_2 + \left(x^2\right)^2 \vec{e}_2\cdot\vec{e}_2 .$$

$$= g_{ij}x^ix^j$$

Defining

$$g_{ij} = \vec{e}_i\cdot\vec{e}_j = \begin{pmatrix} 1 & \cos\alpha \\ \cos\alpha & 1 \end{pmatrix} = g_{ji}$$

$$\therefore |g_{ij}| = \sin^2\alpha.$$

$\therefore$ The inverse of $g_{ij} = \dfrac{1}{\sin^2\alpha}\begin{pmatrix} 1 & -\cos\alpha \\ -\cos\alpha & 1 \end{pmatrix}$.

$\therefore$ If we denote the inverse of g_{ij} as g^{ij}, then $g^{ij} = \dfrac{1}{\sin^2\alpha}\begin{pmatrix} 1 & -\cos\alpha \\ -\cos\alpha & 1 \end{pmatrix}$.

$\therefore$ Equation (1.2.3) can be written as $x_i = g_{ij}x^j$
Otherwise, $x^i = g^{ij}x_j$

so that $g_{ij}\,g^{jk} = \begin{pmatrix} 1 & 0 \\ 0 & 1 \end{pmatrix}$, and hence

$$|\bar{A}|^2 = g_{ij}x^ix^j$$

$$= x_i\, x^i$$

$$= x_i g^{ij} x_j$$

$$= g^{ij} x_i x_j .$$

Definition

Basis: A non-empty subset $S = \{\alpha_1, \alpha_2, \ldots, \alpha_n\}$ of a vector space $V(F)$ is said to be its basis if

i. S is linearly independent of V.
ii. S generates V, i.e., if every vector $\vec{\alpha} \in V$ is expressible in terms of the basis set $\{\alpha_i\}$. But S is not unique for if $S = \{\alpha_1, \alpha_2, \ldots, \alpha_n\}$ is the

basis, then $\{c\alpha_1, \alpha_2, \ldots, \alpha_n\}$ is also the basis. For classical example, any vector $\vec{d}$ in three dimensions can be expressed in terms of three noncoplanar vectors $\bar{a}, \bar{b}, \bar{c}$ as $\vec{d} = \lambda\vec{a} + \mu\vec{b} + \upsilon\vec{c}$.

Symbolically, if $\vec{b}_{(i)}$ are three noncoplanar vectors, they can be taken as the basis of vector $\bar{a}$ so that $\vec{a} = a_i\vec{b}_{(i)}$, where a_i's are the components.

Definition

Coordinate basis: The coordinates of a point in n-dimensional space are identified with reference to a set of axes, and every point x_i $(i = 1, 2, \ldots, n)$ can be correlated with the position vector $\bar{r}(= x_i\hat{e}_i) = x_1\hat{e}_1 + x_2\hat{e}_2 + \cdots + x_n\hat{e}_n)$ where each term represents the displacement (vector) in the direction of the respective axis. The set $\{\hat{e}_i\}$ is called the coordinate basis. For example, in the three-dimensional Euclidean space, $E_3, \vec{r} = \hat{i}x + \hat{j}y + \hat{k}z$ is the position vector of the point (x, y, z) with the coordinate bases $\hat{i}$, $\hat{j}$, and $\hat{k}$.

Orthogonality: Let $\vec{\alpha}$ and $\vec{\beta}$ be any two vectors in an inner product (or dot product) space. The vector $\vec{\alpha}$ is said to be orthogonal to the vector $\vec{\beta}$ if $\vec{\alpha}\cdot\vec{\beta}(=< \vec{\alpha}, \vec{\beta} >) = 0$. If for any two vectors $\vec{\alpha}_i, \vec{\alpha}_j$ $(i \neq j)$ of a set of vectors $S = \{\vec{\alpha}_1, \vec{\alpha}_2, \ldots, \vec{\alpha}_n\}$, $\vec{\alpha}_i \cdot \vec{\alpha}_j = 0(=< \vec{\alpha}_i, \vec{\alpha}_j >)$, then the set is called the orthogonal set.

Orthonormal set: A set $S = \{\hat{\alpha}_1, \hat{\alpha}_2, \ldots, \hat{\alpha}_n\}$ of vectors $\vec{V}(F)$ is said to be orthonormal if

$$\vec{\alpha}_i \cdot \vec{\alpha}_j = 0\,(=< \vec{\alpha}_i, \vec{\alpha}_j >),\ if\ i \neq j$$
$$= 1, \quad if\ i = j.$$

Norm: For any inner product space V, the norm (or length or magnitude) of any vector $\vec{\alpha} \in V$ is defined by $\|\alpha\| = \sqrt{< \bar{\alpha}, \bar{\alpha} >} = \sqrt{\vec{\alpha}\cdot\vec{\alpha}}$ $\vec{\alpha} \neq 0$, and it is a nonnegative value. Multiplying each of the vectors of S by the reciprocal of its norm or length, S can be transformed to an orthonormal set.

N.B.: The coordinate bases may not be orthogonal or orthonormal.

1.3 Quadratic Forms, Properties, and Classifications

Definition

A homogeneous second-degree polynomial in n variables (in general) $x^1, x^2, \ldots, x^n$ is called a quadratic form and is expressed as

$$a_{ij}x^i x^j, \tag{1.3.1}$$

with double sum $i, j = 1, 2, \ldots, n$.

The form is said to be real if all the coefficients a_{ij} are real, and it is called nonsingular if the corresponding determinant

$$|a_{ij}| = \begin{vmatrix} a_{11} & a_{12} & \ldots & a_{1n} \\ a_{21} & a_{22} & \ldots & a_{2n} \\ a_{n1} & a_{n2} & \ldots & a_{nn} \end{vmatrix} \neq 0,$$

Otherwise, it is called singular.

Further, the rank of the square matrix (a_{ij}) represents the rank of the quadratic form.

Again with the help of a nonsingular linear transformation $x^i = \alpha^i_j\, y^j$, the quadratic form can be reduced to the form $c_{ij} y^i y^j (i, j = 1, 2, \ldots, n)$ in terms of the variables y^i subject to the well-established result: rank of the matrix (a_{ij}) = rank of the matrix (c_{ij}).

Theorem

If r is the rank of a real quadratic form $(c_{ij})\, y^i\, y^j$, there exists a nonsingular linear transformation of the variables which can reduce it to the form $c_r (x^r)^2$ where none of c_r's is zero.

The quadratic form can be expressed as

$$c_1 (x^1)^2 + c_2 (x^2)^2 + \cdots + c_r (x^r)^2. \tag{1.3.2}$$

Here, c_r can be a positive or negative nonzero constant.

Signature: The difference between the number of positive and negative coefficients or the excess of positive coefficients over the negative coefficients of (1.3.2) is called the **signature** of the real quadratic forms.

Of course, according to the "Sylvester's law of inertia," the number of positive coefficients is invariant (remains unchanged).

Normal form: If the real quadratic form $c_r (x^r)^2$ is transformable by means of a nonsingular linear transformation to a form in which the coefficients c_r's take up values +1 and –1, it is called the normal form of it.

If p is the number of positive coefficients (+1), q is the number of negative coefficients (–1), and S is the signature, then the normal form can be written as

$$\left(x^1\right)^2 + \left(x^2\right)^2 + \cdots + \left(x^p\right)^2 - \left(x^{p+1}\right)^2 - \left(x^{p+2}\right)^2 - \cdots - \left(x^r\right)^2, \tag{1.3.3}$$

where $p + q = r$ and $p - q = s$, so that $p = \frac{1}{2}(r + s)$.

Definite and indefinite quadratic forms: If all the signs of the quadratic normal form (1.3.3) are the same (positive or negative) or different, then the quadratic form is called **definite** or **indefinite**.

Positive definite and negative definite quadratic forms: If all the signs of the normal quadratic form (1.3.3) are positive, it is called positive definite, and if all of them are negative, it is called negative definite.

1.4 Quadratic Differential Forms and Metric of a Space in the Form of Quadratic Differentials

Definition

A second-degree homogeneous polynomial of the differentials dx^i of the variables $x^1, x^2, \ldots, x^n$ is called a quadratic differential form, e.g., $Q(a) = a_{ij}\, dx^i\, dx^j$ is a quadratic differential form where a_{ij}'s are the functions of x^i's or may be constant. A quadratic differential form is of paramount importance, which will be demonstrated in the next few chapters.

Considering n independent functions $y^i = y^i(x^1, x^2, \ldots, x^n)$ instead of the variables x^i, we can get a homogeneous linear transformation of differentials:

$$dy^i = \frac{\partial y^i}{\partial x^j} dx^j. \tag{1.4.1}$$

This can also be written as $dx^i = \dfrac{\partial x^i}{\partial y^j} dy^j$, since for nonsingular linear transformation of quadratic forms of the same rank $\left|\dfrac{\partial y^i}{\partial x^j}\right| \neq 0$.

$\therefore$ The quadratic differential form $a_{ij}\, dx^i dx^j$ can be transformed to

$$Q(b) = b_{ij}\, dy^i\, dy^j. \tag{1.4.2}$$

Noticeably at a **given point** or for a **given** x^i, the relation (1.4.1) stands for a linear transformation of differentials with constant coefficient. Hence, it is analogous to quadratic forms discussed in Section 1.3. Contextually, (1.4.2) can also be reduced to the similar normal form:

$$\left(dy^1\right)^2 + \left(dy^2\right)^2 + \cdots + \left(dy^p\right)^2 - \left(dy^{p+1}\right)^2 - \left(dy^{p+2}\right)^2 - \cdots - \left(dy^r\right)^2, \tag{1.4.3}$$

where r is the rank of (b_{ij}).

The notions called positive definite, negative definite, and signature of the normal form of this quadratic differential are exactly similar to those as discussed in Section 1.3.

For simplicity, let us consider the position vector $\bar{r}$ of a point P (as in Figure 1.1) in curvilinear coordinates u_1, u_2, u_3 as $\bar{r} = \bar{r}(u_1, u_2, u_3)$ in three dimensions (Figure 1.2).

$$\therefore\ d\bar{r} = \frac{\partial \bar{r}}{\partial u_1} du_1 + \frac{\partial \bar{r}}{\partial u_2} du_2 + \frac{\partial \bar{r}}{\partial u_3} du_3.$$

If ds is the element (length) between the adjacent points $P(\bar{r})$ and $Q(\bar{r} + d\bar{r})$, then (in the limit)

$$d\bar{r} \cdot d\bar{r} = ds^2$$

$$= \left(\frac{\partial \bar{r}}{\partial u_1} du_1 + \frac{\partial \bar{r}}{\partial u_2} du_2 + \frac{\partial \bar{r}}{\partial u_3} du_3\right) . \left(\frac{\partial \bar{r}}{\partial u_1} du_1 + \frac{\partial \bar{r}}{\partial u_2} du_2 + \frac{\partial \bar{r}}{\partial u_3} du_3\right)$$

$$= \left(\frac{\partial \bar{r}}{\partial u_1}\right)^2 du_1^2 + 2\left(\frac{\partial \bar{r}}{\partial u_1} . \frac{\partial \bar{r}}{\partial u_2}\right) du_1\, du_2 + 2\left(\frac{\partial \bar{r}}{\partial u_1} . \frac{\partial \bar{r}}{\partial u_3}\right) du_1\, du_3$$

$$+ \left(\frac{\partial \bar{r}}{\partial u_2}\right)^2 du_2^2 + 2\left(\frac{\partial \bar{r}}{\partial u_2} . \frac{\partial \bar{r}}{\partial u_3}\right) du_2\, du_3 + \left(\frac{\partial \bar{r}}{\partial u_3}\right)^2 du_3^2$$

$$= a_{11}\, du_1^2 + 2a_{12} du_1\, du_2 + 2a_{13} du_1\, du_3 + a_{22}\, du_2^2 + 2a_{23} du_2\, du_3 + a_{33} du_3^2$$

$$\therefore ds^2 = \sum_{i=1}^{3} \sum_{j=1}^{3} a_{ij}\, du_i\, du_j,$$

where

$$a_{ij} = \frac{\partial \bar{r}}{\partial u_i} . \frac{\partial \bar{r}}{\partial u_j} \neq 0$$

and

$$ds^2 = a_{ij}\, du_i\, du_j \tag{1.4.4}$$

by dropping summation sign for summation convention (defined in Section 2.1).

This is the quadratic differential form representing the elementary distance between two adjacent points in the oblique curvilinear coordinate system. This is called the metric or line element of the system. Of course, in oblique curvilinear coordinates (u, v, w), (1.4.4) can be written in explicit form as

$$ds^2 = adu^2 + bdv^2 + cdw^2 + 2h\, dudv + 2gdudw + 2fdvdw.$$

But in rectangular Cartesian coordinates of Euclidean space of three dimensions, the elementary distance between the adjacent points (x, y, z) and $(x+dx, y+dy, z+dz)$ is given by

$$ds^2 = dx^2 + dy^2 + dz^2,$$

where

$$a_{ij} = \frac{\partial \bar{r}}{\partial u_i} \cdot \frac{\partial \bar{r}}{\partial u_j} = 0 \text{ for } i \neq j$$

$$= 1 \text{ for } i = j$$

$$u_1 = x, u_2 = y, u_3 = z.$$

To include almost all physical spaces, Riemann has generalized this concept (notion) to n dimensions. This will be discussed in detail in Chapter 3.

Exercises

1. If the quadratic form $a_{ij}x^i x^j$ transforms to a quadratic form $b_{ij}y^i y^j$, write down the corresponding form of nonsingular linear transformation of it and what is the rank of $|b_{ij}|$ if the rank of $|a_{ij}|$ is 5 under suitable transformation in the range of i and j.
2. If the quadratic form $a_{ij}x^i x^j$ $(i, j = 1, 2, \ldots, n)$ reduces to the form $b_i(y^i)^2$, $b_i \neq 0 \ \forall\, i$ by means of a nonsingular linear transformation, what are the nonzero values of b's if r $(<n)$ is the rank of the latter.
3. With the help of an example, show that the signature of a real quadratic form remains the same.
4. Show that a definite quadratic form cannot be singular.
5. What are the loci (if possible) represented by the quadratic form $a_i x^i x^i = 1$, where $N = 2, 3$ where $x^i = 1, 2, \ldots, N$ are rectangular coordinates.

Notes

2

Concept of Tensors

2.1 Some Useful Definitions

In mathematics, symbolic representations of some concepts are essential for writing expressions in concise form for convenience. For example,

Summation convention: It means the drop of summation sign (Σ) in an expression of the type $\sum_{i=1}^{n} a_i x^i = a_i x^i$. It's explicit form is $a_i x^i = a_1 x^i + a_2 x^2 + \cdots + a_n x^n$, a linear expression in terms of n variables $x^1, x^2, \ldots, x^n$; the index does not mean power rather designates variable (position).

Dummy index: An index that occurs twice in a term once in the upper position (prefix) and once in the lower position (subscripts) is called a dummy index, e.g., in $a_i x^i$, i is the dummy index which indicates the summation of the desired range, say, from 1 to n (if not mentioned).

Illustration: In order to write a system of linear equations of n unknowns, we can use $a_i^\alpha x^i = b^\alpha$, where $\alpha = 1, 2, \ldots, n$, $i = 1, 2, \ldots, n$. Actually, it stands for

$$a_1^1 x^1 + a_2^1 x^2 + \cdots + a_n^1 x^n = b^1$$

$$a_1^2 x^1 + a_2^2 x^2 + \cdots + a_n^2 x^n = b^2$$

$$\vdots$$

$$a_1^n x^1 + a_2^n x^2 + \cdots + a_n^n x^n = b^n.$$

These n linear equations in n unknowns are **concisely** written as $a_i^\alpha x^i = b^\alpha$.

This is definitely a stylistic representation of mathematics.

Real index: An index that occurs only once in a term like α in $a_i^\alpha x^i = b^\alpha$ is called a real index.

N.B.: A dummy index can be replaced by any other index not occurring in that term, e.g., $a_i x^i = a_j x^j$ indicating the same expression. But a real index cannot be replaced except on either sides.

Kronecker delta: Let x^i ($i = 1, 2, \ldots, n$) be a set of n **independent** variables. Then,

$$\frac{\partial x^i}{\partial x^j} = 1, \text{ when } i = j$$

$$= 0, \text{ when } i \neq j$$

It is denoted by the symbol δ^i_j called the Kronecker delta.

$$\therefore \left(\frac{\partial x^i}{\partial x^j} = \right)\delta^i_j = 1, \text{ when } i = j$$

$$= 0, \text{ when } i \neq j,$$

2.2 Transformation of Coordinates

Let us consider the set of n real independent variables $x^i (i = 1, 2, \ldots, n)$ to denote the coordinates of a point P with respect to a system x^i in an n-dimensional space V_n. Otherwise, every n tuple of real variables corresponds to a point in a space V_n. Let $x'^i (i = 1, 2, \ldots, n)$ be the coordinates of the same point P but with respect to another system x'^i (say). Since the **point is the same,** the two sets of variables must be functionally connected by means of a transformation as $x'^i = f(x^i)$.

From differential calculus, we can get

$$dx'^i = \frac{\partial f}{\partial x^1} dx^1 + \frac{\partial f}{\partial x^2} dx^2 + \cdots + \frac{\partial f}{\partial x^n} dx^n$$

$$= \sum_{\alpha=1}^{n} \frac{\partial f}{\partial x^\alpha} dx^\alpha = \sum \frac{\partial x'^i}{\partial x^\alpha} dx^\alpha$$

or

$$dx'^i = \frac{\partial x'^i}{\partial x^\alpha} dx^\alpha \qquad (2.2.1)$$

by summation convention, where $\left|\frac{\partial x'^i}{\partial x^\alpha}\right|$ is the nonzero Jacobian of the transformation.

The infinitesimal displacement dx^i for the coordinate system x^i determines the **direction at** P in V_n, and they are called the components of a contravariant vector; dx'^i are the corresponding contravariant vector components in the other system x'^i.

Defining $dx^\alpha = A^\alpha$, (2.2.1) can be written as

$$A'^i = \frac{\partial x'^i}{\partial x^\alpha} A^\alpha. \tag{2.2.2}$$

Thus, if the set of n functions $A^\alpha(\alpha = 1,2,\ldots,n)$ of the x^i coordinate system transforms to the set of n functions $A'^i(i = 1,2\ldots,n)$ in another system x'^i by means of the transformation law (2.2.2), it is called the components of a contravariant vector.

Multiplying (2.2.2) by $\frac{\partial x^\beta}{\partial x'^i}$ and summing over i, we can get

$$\frac{\partial x^\beta}{\partial x'^i} A'^i = \frac{\partial x^\beta}{\partial x'^i}\frac{\partial x'^i}{\partial x^\alpha} A^\alpha = \frac{\partial x^\beta}{\partial x^\alpha} A^\alpha = \delta^\beta_\alpha A^\alpha$$

$\therefore A^\beta = \frac{\partial x^\beta}{\partial x'^i} A'^i$, which is the reverse transformation law of (2.2.2).

Again from the transformation $x'^i = f(x^i)$, we can solve for x^i in terms x'^i of another function $\phi(x'^i)$ as the Jacobian of the transformation $\left|\frac{\partial x'^i}{\partial x^j}\right| \neq 0$. In this sense, $x^i = \phi(x'^i)$.

Let the function $\phi(x'^i)$ be invariant subject to coordinate transformation (in general) or in restricted sense be a scalar invariant. Therefore, its partial derivatives with respect to the coordinates x'^i, namely $\frac{\partial \phi}{\partial x'^j}$, admit the coordinate transformation according to the rule:

$$\frac{\partial \phi}{\partial x'^i} = \frac{\partial \phi}{\partial x^1}\cdot\frac{\partial x^1}{\partial x'^i} + \frac{\partial \phi}{\partial x^2}\frac{\partial x^2}{\partial x'^i} + \cdots + \frac{\partial \phi}{\partial x^n}\frac{\partial x^n}{\partial x'^i}$$

$$= \frac{\partial \phi}{\partial x^j}\frac{\partial x^j}{\partial x'^i}(j = 1,2,\ldots,n),$$

by summation convention.

Putting $\frac{\partial \phi}{\partial x^j} = A_j$, it can be thrown to the form:

$$A'_i = \frac{\partial x^j}{\partial x'^i} A_j \tag{2.2.3}$$

where $A_j(j = 1,2,\ldots,n) = \frac{\partial \phi}{\partial x^j}$; the functions of the coordinates x^j are called the components of a covariant vector, and A'_j are its corresponding components in the x'^i coordinate system. Otherwise, any set of n functions $A_j (j=1,2,\ldots,n)$ of the x^i coordinate system is transformed to another set of n functions $A'_i (i=1,2,\ldots,n)$ of the x'^i coordinate system according to the transformation law (2.2.3) is called the components of a covariant vector.

The partial derivatives $\frac{\partial \phi}{\partial x^j}$ are the components of grad ϕ vector and if $\phi(x^i) = \text{const}(i = 1,2,3 \text{ say})$ like invariant, then $\nabla\phi$ represents a vector normal to the surface $\phi = c$ in three dimensions.

N.B.: The readers are advised to delve in coining the names "contravariant vector" and "covariant vector" in the above definitions (2.2.2) and (2.2.3) with the earlier uses of the definitions discussed in Section 1.2.

Cartesian tensors: If the components of a tensor are expressed relative to Cartesian coordinates and the operative transformation is from Cartesian to Cartesian, they are called **Cartesian tensors.**

Theory of continuum mechanics is developed on the basis of Cartesian tensors.

Definition

Relative and absolute tensors: If the components of a tensor (may be of any order, see Section 2.3) $A^{ijk}_{\alpha\beta}$ are transformed according to the law,

$$A'^{mnr}_{st} = \left(\frac{\partial x}{\partial x'}\right)^w A^{ijk}_{\alpha\beta} \frac{\partial x'^m}{\partial x^i} \frac{\partial x'^n}{\partial x^j} \frac{\partial x'^r}{\partial x^k} \frac{\partial x^\alpha}{\partial x'^s} \frac{\partial x^\beta}{\partial x'^t}$$

where $\left|\frac{\partial x}{\partial x'}\right|$ is the Jacobian of the transformation associated with the factor $\left(\frac{\partial x}{\partial x'}\right)^w$, then the tensors are called relative tensors of **weight** w.

When $w = 0$, the tensor is called absolute tensor, and when $w = 1$, the relative tensor is called a **tensor density.**

Definition

Permutation (or pseudo) tensor: The absolute tensor $\in_{jik}$ is defined as follows:

$\in_{jik} = 0$, if any two of the indices i, j, k repeat
$= 1$, if i, j, k is an **even** permutation of 1, 2, 3
$= -1$, if i, j, k is an **odd** permutation of 1, 2, 3 is called permutation or pseudo tensor.

Observation: According to the transformation law (2.2.2) for the two sets of functions A^α and A'^i, we can write

$$A'^i = \frac{\partial x'^i}{\partial x^\alpha} A^\alpha \text{ and } B'^j = \frac{\partial x'^j}{\partial x^\beta} B^\beta,$$

so that

$$A'^i B'^j = A^\alpha \frac{\partial x'^i}{\partial x^\alpha}\frac{\partial x'^j}{\partial x^\beta} B^\beta = (A^\alpha B^\beta)\frac{\partial x'^i}{\partial x^\alpha}\frac{\partial x'^j}{\partial x^\beta}$$

$$C'^{ij} = C^{\alpha\beta}\frac{\partial x'^i}{\partial x^\alpha}\frac{\partial x'^j}{\partial x^\beta} \quad \text{if we recognize } A^\alpha B^\beta = C^{\alpha\beta}.$$

This prompts us to define the product of two vectors.

2.3 Second and Higher Order Tensors

Any set of n^2 quantities $A^{ij}(i, j = 1, 2, \ldots, n)$ is called a second-order (rank) contravariant tensor if it satisfies the transformation law:

$$A'^{ij} = \frac{\partial x'^i}{\partial x^a}\frac{\partial x'^j}{\partial x^b} A^{ab}. \tag{2.3.1}$$

A set of n^2 quantities A^{ij} $(i, j = 1, 2, \ldots, n)$ in the x^i coordinate system transformed to another set of n^2 quantities A'^{ij} in the x'^i coordinate system according to the transformation law (2.3.1) is called the contravariant tensor of second order (or of rank two).

Similarly, if a set of n^2 quantities B_{ij} $(i, j = 1, 2, \ldots, n)$ in the x^i coordinate system transforms to another set of n^2 quantities B'_{ij} in the x'^i coordinate system according to the transformation law,

$$B'_{ij} = \frac{\partial x^a}{\partial x'^i}\frac{\partial x^b}{\partial x'^j} B_{ab} \tag{2.3.2}$$

is called the covariant tensor of second order (or of rank two).

Also, if a tensor of the type A^i_j covariant character by one and contravariant character by one in a coordinate system x^i transforms to A'^i_j in the x'^i coordinate system according to the law of transformation,

$$A'^i_j = \frac{\partial x'^i}{\partial x^a}\frac{\partial x^b}{\partial x'^j} A^a_b$$

is called a mixed tensor of second order.

Definition

Order or rank of a tensor: The number of designated real index in a tensor is called the rank or order of a tensor. A'^{ij}, A_{ijk} are the tensors of rank two and three, respectively, of contravariant and covariant characters.

i. A vector is a tensor of rank one with $3^1(\text{or } n^1)$ components in three dimensions (n dimensions).
ii. A stress α_{ij} is a tensor of rank two with $3^2 = 9(\text{or } n^2)$ components in three dimensions (n dimensions).
iii. An invariant is also called a tensor of rank (or order) zero.
iv. The upper (superscript) index is reserved to represent contravariant character and lower (subscript) to represent covariant character.

2.4 Operations on Tensors

Definition

Outer or open product of tensors: The components of a tensor of any rank (or order) and character can be multiplied with the components of a tensor of any rank and character, and the rank (order) of the resulting tensor is the sum of their individual ranks.

For example, the product of the **second**-order mixed tensor A_j^i with the **third**-order contravariant tensor B^{mnr} is C_j^{imnr}, a fifth-order mixed tensor.

This is called **outer or open product** of two tensors.

Contraction: The process of equating a contravariant index with a covariant index in a mixed tensor implicating summation with respect to it is called a contraction.

For example, the tensor A_{ab}^{ijk} may be subjected to several contractions, namely,

$$A_{ib}^{ijk}, A_{ai}^{ijk}, A_{jb}^{ijk}, A_{aj}^{ijk}, A_{kb}^{ijk}, A_{ak}^{ijk}.$$

In all these tensors, the number of real indices transforms to three instead of five as in the original tensor. Can it not be guessed the following?

Theorem

Every contraction reduces the rank of a tensor by two.

Proof: Let us consider a fifth-order mixed tensor A_{qp}^{ijk} in the x^i system. By transformation law $A_{st}'^{mnr}$ in the x'^i system, we can write

$$A'^{mnr}_{st} = \frac{\partial x'^m}{\partial x^i}\frac{\partial x'^n}{\partial x^j}\frac{\partial x'^r}{\partial x^k}\frac{\partial x^q}{\partial x'^s}\frac{\partial x^p}{\partial x'^t}A^{ijk}_{qp}$$

$$A'^{mnr}_{sr} = \frac{\partial x'^m}{\partial x^i}\frac{\partial x'^n}{\partial x^j}\frac{\partial x'^r}{\partial x^k}\frac{\partial x^q}{\partial x'^s}\frac{\partial x^p}{\partial x'^r}A^{ijk}_{qp}$$

taking $r = t$, i.e., assuming a contraction

$$A'^{mnr}_{sr} = \frac{\partial x'^m}{\partial x^i}\frac{\partial x'^n}{\partial x^j}\frac{\partial x^p}{\partial x^k}\frac{\partial x^q}{\partial x'^s}A^{ijk}_{qp}$$

$$= \frac{\partial x'^m}{\partial x^i}\frac{\partial x'^n}{\partial x^j}\delta^p_k\frac{\partial x^q}{\partial x'^s}A^{ijk}_{qp}$$

$$A'^{mnr}_{sr} = \frac{\partial x'^m}{\partial x^i}\frac{\partial x'^n}{\partial x^j}\frac{\partial x^q}{\partial x'^s}1\cdot A^{ijk}_{qk}$$

which is the transformation law of a third-order mixed tensor as the dummy index r on the left and k on the right are not countable in the rank.

Hence, proved.

Inner product of tensors: A multiplication associated with a contraction in tensors is called an inner product of tensors.

For example, $A_{ij}B^{jkr} = C_i^{kr}$ is a third-order tensor instead of the fifth-order tensor due to one contraction.

$$A_{ij}B^{jir} = C^r$$

is the first-order tensor (vector) due to two successive contractions.

Higher order tensors: After formulation of second-order tensors and multiplication (outer or open product) of tensors, we can define tensors of any order.

The mth-order contravariant tensor $A^{r_1 r_2,\ldots,r_m}$ connecting the n^m components in the x^i system with the corresponding n^m components in the x'^i system can be defined as

$$A'^{s_1,s_2,\ldots,s_m} = \frac{\partial x'^{s_1}}{\partial x^{r_1}}\frac{\partial x'^{s_2}}{\partial x^{r_2}}\cdots\frac{\partial x'^{s_m}}{\partial x^{r_m}}A^{r_1,r_2,\ldots,r_m}$$

Similarly, a kth-order covariant tensor $A_{abc\ldots s}$ of n^k components in the x^i system can be defined by means of the transformation to the corresponding n^k components in the x'^i system as

$$A'_{pqr\ldots u} = \frac{\partial x^a}{\partial x'^p}\frac{\partial x^b}{\partial x'^q}\frac{\partial x^c}{\partial x'^r}\cdots\frac{\partial x^s}{\partial x'^u}A_{abc\ldots s},$$

where $(p, q, r, \ldots, u)$ or $(a, b, c, \ldots, s)$ is the k number of indices. Any mixed tensor of any order can similarly be defined.

Addition (subtraction) of tensors: Two tensors of the same rank and similar character (covariant or contravariant) can be added (components), and the components of one can be subtracted from the other. Otherwise, the sum or difference of two tensors of the same rank and character is a tensor of the same type. For example,

Let A^i_{jk} and B^i_{jk} be the two third-order mixed tensors of the same character so that

$$A^i_{jk} + B^i_{jk} = C^i_{jk}, \text{(say) where } C^i_{jk} \text{ is doubtful.} \tag{2.4.1}$$

In the x'^i system, its replica is

$$C'^m_{nr} = A'^m_{nr} + B'^m_{nr} = \frac{\partial x'^m}{\partial x^i}\frac{\partial x^j}{\partial x'^n}\frac{\partial x^k}{\partial x'^r} A^i_{jk} + \frac{\partial x'^m}{\partial x^i}\frac{\partial x^j}{\partial x'^n}\frac{\partial x^k}{\partial x'^r} B^i_{jk}$$

$$= \left(A^i_{jk} + B^i_{jk}\right)\frac{\partial x'^m}{\partial x^i}\frac{\partial x^j}{\partial x'^n}\frac{\partial x^k}{\partial x'^r} = C^i_{jk}\frac{\partial x'^m}{\partial x^i}\frac{\partial x^j}{\partial x'^n}\frac{\partial x^k}{\partial x'^r},$$

(using 2.4.1) which is the transformation law of a third-order mixed tensor of the same character.

Hence, C^i_{jk} (the doubtful one) is also a tensor of the same kind.

Similarly, the difference of two tensors of the same type like $A^i_{jk} - B^i_{jk}$(or $A^{ij} - B^{ij}$) can also be proved to be the tensor of the same type.

2.5 Symmetric and Antisymmetric (or Skew-Symmetric) Tensors

When the relative position of two indices contravariant or covariant in the components is interchangeable in a tensor, it is called a symmetric tensor with respect to these indices, i.e., A^i_{jk} (or A^{ij}) is symmetric if $A^i_{jk} = A^i_{kj}$ (or $A^{ij} = A^{ji}$).

Similarly, A_{ij} or A^{ij} is said to be antisymmetric (or skew-symmetric) if

$$A_{ij} = -A_{ji} \text{ or } A^{ij} = -A^{ji}.$$

Theorem

The symmetric (or antisymmetric) property of tensors remains unchanged (invariant) due to transformation of coordinates.

Proof: Let A_{ij} be the components of a symmetric tensor in the x^i coordinate system so that $A_{ij} = A_{ji}$.

If A'_{mn} are the corresponding components of the tensor in the x'^i coordinate system, then it is to satisfy

$$A'_{mn} = \frac{\partial x^i}{\partial x'^m}\frac{\partial x^j}{\partial x'^n}A_{ij}$$

$$= \frac{\partial x^j}{\partial x'^n}\frac{\partial x^i}{\partial x'^m}A_{ji}.$$

$$\therefore A_{ij} = A_{ji}$$

$$A'_{mn} = A'_{nm}$$

Hence, proved.

Similarly, it can also be proved that

$$A'^{ij} = -A'^{ji} \text{ if } A^{ij} = -A^{ji}.$$

Example 1

Prove that every second-order tensor covariant (or contravariant) can be expressed as the sum of symmetric and antisymmetric tensors.

Let A_{ij} be any second-order covariant tensor. It can be written as

$$A_{ij} = \frac{1}{2}(A_{ij} + A_{ji}) + \frac{1}{2}(A_{ij} - A_{ji})$$

$$\therefore A_{ij} = B_{ij} + C_{ij}, \quad \text{(i)}$$

where

$$\frac{1}{2}(A_{ij} + A_{ji}) = B_{ij} \quad \text{(ii)}$$

$$\frac{1}{2}(A_{ij} - A_{ji}) = C_{ij} \quad \text{(iii)}$$

as they are conformable for addition and subtraction of tensors.

In view of writing (ii),

$$B_{ji} = \frac{1}{2}(A_{ji} + A_{ij}) = \frac{1}{2}(A_{ij} + A_{ji})$$

$$\therefore B_{ij} = B_{ji}.$$

Hence, B_{ij} is a symmetric tensor.

Also, in view of writing (iii),

$$C_{ji} = \frac{1}{2}\left(A_{ji} - A_{ij}\right) = -\frac{1}{2}\left(A_{ij} - A_{ji}\right) = -C_{ij}.$$

Hence, C_{ji} is an antisymmetric tensor.

From (i), A_{ij} is expressed as the sum of symmetric and antisymmetric tensors.

Similarly, it can be proved for second-order contravariant tensor A^{ij} also.

N.B.: This result is similar to the result of a square matrix, namely, every square matrix can be expressed as the sum of symmetric and antisymmetric matrices.

Example 2

If A^i are the components of any contravariant vector in the sum A^iB_i which is a scalar invariant, prove that B_i are also the components of a covariant vector.

Given A^iB_i is a scalar invariant.

$\therefore$ For a transformation of coordinates from the x^i system to the x'^i system,

$$A'^iB'_i = A^iB_i \quad \therefore i \text{ is dummy}$$

$$\frac{\partial x'^i}{\partial x^\alpha} A^\alpha B'_i = A^\alpha B_\alpha \quad (i \to \alpha)$$

$$\therefore A^\alpha \left(B_\alpha - \frac{\partial x'^i}{\partial x^\alpha} B'_i \right) = 0$$

$$B_\alpha = \frac{\partial x'^i}{\partial x^\alpha} B'_i.$$

$\because A^\alpha$ is any contravariant vector.

This is the transformation law of covariant vector.

Hence, B_α or B_i is also a covariant vector.

N.B.: It will be seen later that A^iB_i is the scalar product of two vectors A^i and B_i.

Example 3

The number of independent components of a second-order symmetric tensor in n dimensions is $\frac{1}{2}\, n(n+1)$.

The total number of components of a second-order covariant tenser A_{ij} (say) is n^2. But the number of components with repeated suffix A_{ii} is n.

Hence, the number of components with distinct indices A_{ij} $(i \neq j)$ is $n^2 - n$. But for symmetric property $A_{ij} = A_{ji}$, the actual number of independent components out of these is $\frac{1}{2}\,(n^2 - n)$.

$\therefore$ Total number of independent components is

$$= \frac{1}{2}(n^2 - n) + n = \frac{1}{2}n^2 + \frac{1}{2}n = \frac{1}{2}n(n+1)$$

Hence, proved.

Example 4

If n^3 quantities B^i_{jk} are connected with n^3 quantities A^i_{jk} by the relations, $B^i_{jk} = A^h_{ml} \frac{\partial y^i}{\partial x^h} \frac{\partial x^m}{\partial y^j} \frac{\partial x^l}{\partial y^k}$, show that $B^i_{ik} = A^i_{ij} \frac{\partial x^l}{\partial y^k}$.

Given $B^i_{jk} = A^h_{ml} \frac{\partial y^i}{\partial x^h} \frac{\partial x^m}{\partial y^j} \frac{\partial x^l}{\partial y^k}$.

Putting $i = j$, i.e., making a contraction with respect to i and j,

$$B^i_{ik} = A^h_{ml} \left(\frac{\partial y^i}{\partial x^h} \frac{\partial x^m}{\partial y^i} \right) \frac{\partial x^l}{\partial y^k} = A^h_{ml} \frac{\partial x^m}{\partial x^h} \frac{\partial x^l}{\partial y^k}$$

$$= A^h_{ml} \delta^m_h \frac{\partial x^l}{\partial y^k} = A^h_{hl} \frac{\partial x^l}{\partial y^k} = A^i_{il} \frac{\partial x^l}{\partial y^k} \quad (h \to i).$$

Hence, proved.

Example 5

If a_{ij} and b_{ij} are the two symmetric tensors, and u^i, v^i are the components of contravariant vectors satisfying the equations

$$\left(a_{ij} - kb_{ij}\right)u^i = 0 \text{ and } \left(a_{ij} - k'b_{ij}\right)v^i = 0,$$

$$i, j = 1, 2, \ldots, n \quad k \neq k'$$

prove that $b_{ij}\, u^i v^j = 0$ and $a_{ij}\, u^i v^j = 0$.

Given $a_{ij} = a_{ji}$, $b_{ij} = b_{ji}$ and

$$(a_{ij} - kb_{ij})u^i = 0 \tag{i}$$

and

$$(a_{ij} - k'b_{ij})v^i = 0$$

$$(a_{ji} - k'b_{ji})v^j = 0. \tag{ii}$$

Multiplying (i) by v^j and summing over j and multiplying (ii) by u^i and summing over i from $i, j = 1, 2, \ldots, n$, we get

$$(a_{ij} - kb_{ij})u^i v^j = 0 \tag{iii}$$

$$(a_{ji} - k'b_{ji})u^i v^j = 0. \qquad \text{(iv)}$$

(iii)–(iv) gives

$$(k - k')b_{ij}u^i v^j = 0 \quad \therefore a_{ij} = a_{ji} \text{ and } b_{ij} = b_{ji}$$

$$\therefore b_{ij}\ u^i v^j = 0 \qquad k' \neq k.$$

Also (iii) $\times k'$ and (iv) $\times k$ gives

$$(k' - \text{k})a_{ij}\ u^i v^j = 0 \quad \therefore a_{ij}\ u^i v^j = 0 \quad k' \neq k$$

Hence, proved.

Theorem

Show that the tensor law of transformation possesses the group property.

Proof: Consider the contravariant (may be covariant also) components A^i and A'^i of the same vector in x^i and x'^i systems, respectively.

$$\therefore A'^i = \frac{\partial x'^i}{\partial x^\alpha} A^\alpha \qquad \text{(i)}$$

Let us consider a third coordinate system x''^j with components A''^j of the same vector.

$$\therefore A''^j = \frac{\partial x''^j}{\partial x'^i} A'^i = \frac{\partial x''^j}{\partial x'^i}\frac{\partial x'^i}{\partial x^\alpha} A^\alpha,$$

using (i) and summing over i.

$$A''^j = \frac{\partial x''^j}{\partial x^\alpha} A^\alpha$$

which is the direct transformation of tensor (here vector) components from the x^i system to the x''^j system. Thus,

i. Two successive tensor law of transformations give rise to another tensor law of transformation, i.e., identical to closure property of group.
ii. Compositions of tensors are always conformable with respect to addition and multiplication and so associatively is obvious.
iii. Setting $x'^i = x^\alpha$ in (i), we can recover the **same tensor**, namely, $A'^i = A^\alpha$, which is equivalent to the existence of identity.

iv. For nonzero value of the Jacobian $\left|\frac{\partial x'^j}{\partial x^\alpha}\right| \neq 0$, every tensor law of transformation admits the "inverse" transformation (like $A^\alpha = \frac{\partial x^\alpha}{\partial x'^i} A'^i$) law.

Hence, tensor law of transformations are said to possess the group property.

2.6 Quotient Law

If the functions $u^{ijk}_{a\alpha}$ in the x^i system and the functions $u'^{ijk}_{a\alpha}$ in the x'^i system are such that $u^{ijk}_{a\alpha} v^\alpha$ and $u'^{ijk}_{a\alpha} v'^\alpha$ are the components of a tensor, where v^α and v'^α are the components of arbitrary vectors in the systems, then the given functions must be the components of a tensor.

Proof: Given

$$u^{ijk}_{a\alpha} v^\alpha = a \text{ tensor } A^{ijk}_a \text{(say, a tensor in the } x^i \text{ system)} \quad \text{(i)}$$

$$u'^{ijk}_{a\alpha} v'^\alpha = a \text{ tensor } A'^{ijk}_a \left(\text{in the } x'^i \text{ system}\right) \quad \text{(ii)}$$

$$\therefore u'^{ijk}_{a\alpha} v'^\alpha = \frac{\partial x'^i}{\partial x^b}\frac{\partial x'^j}{\partial x^c}\frac{\partial x'^k}{\partial x^d}\frac{\partial x^m}{\partial x'^a} A^{bcd}_m \text{, from (ii)}$$

$$= \frac{\partial x'^i}{\partial x^b}\frac{\partial x'^j}{\partial x^c}\frac{\partial x'^k}{\partial x^d}\frac{\partial x^m}{\partial x'^a}, \quad u^{bcd}_{mn} v^n \text{, using (i)}$$

$$= \frac{\partial x'^i}{\partial x^b}\frac{\partial x'^j}{\partial x^c}\frac{\partial x'^k}{\partial x^d}\frac{\partial x^m}{\partial x'^a} u^{bcd}_{mn} \frac{\partial x^n}{\partial x'^\alpha} v'^\alpha$$

$\because v^n$ (or v^α) is a tensor.

or

$$\left(u'^{ijk}_{a\alpha} - \frac{\partial x'^i}{\partial x^b}\frac{\partial x'^j}{\partial x^c}\frac{\partial x'^k}{\partial x^d}\frac{\partial x^m}{\partial x'^a}\frac{\partial x^n}{\partial x'^\alpha} u^{bcd}_{mn}\right) v'^\alpha = 0$$

$$\therefore u'^{ijk}_{a\alpha} = \frac{\partial x'^i}{\partial x^b}\frac{\partial x'^j}{\partial x^c}\frac{\partial x'^k}{\partial x^d}\frac{\partial x^m}{\partial x'^a}\frac{\partial x^n}{\partial x'^\alpha} u^{bcd}_{mn}.$$

$\because v^\alpha$ (or v'^α) is arbitrary.

This is the transformation law of a fifth-order mixed tensor. Hence, the given functions are also the components of a tensor.

Hence, proved.

N.B.: Some authors state the quotient law as "If the inner product of a class of functions with an arbitrary vector is a tensor, then the class of functions is also a tensor." Students are advised to explore the fallacy in the statement.

Exercises

1. Show that subject to rectangular Cartesian coordinate transformation, there is no distinction between contravariant and covariant vectors (or tensors).
2. From the transformation law of second-order contravariant and covariant tensors, show that $A'^{ij}\dfrac{\partial x^k}{\partial x'^i} = A^{kl}\dfrac{\partial x'^j}{\partial x^l}$ and $A'_{ij}\dfrac{\partial x'^i}{\partial x^k} = A_{kl}\dfrac{\partial x^l}{\partial x'^j}$. [Hint: They can be obtained from the tensor law of transformations.]
3. If the equation $fA_{ij} + gA_{ji} = 0$ is satisfied by the nonzero second-order tensor A_{ij} with respect to a basis, prove that
   i. $f = g$ when $A_{ij} = -A_{ji}$
   ii. $f = -g$ when $A_{ij} = A_{ji}$.
4. If the components of a tensor (of any rank) vanish in one coordinate system, prove that they identically vanish in any other coordinate system. [Hint: Take $A^{ij} = 0$ and write the tensor law of transformation in the x'^i system.]
5. If the relation $a_{ij}b_k + a_{jk}b_i + a_{ki}b_j = 0$ is satisfied by the symmetric tensor a_{ij} and the arbitrary tensor b_k, prove that either $a_{ij} = 0$ or $b_k = 0$.
6. If y^i are n independent functions of the variables x^i, and z^i are n independent functions of the y^i so that $u^i = v^j\dfrac{\partial x^i}{\partial y^j}, v^i = w^j\dfrac{\partial y^i}{\partial z^j}, U_i = V_j\dfrac{\partial y^j}{\partial x^i}$, and $V_i = W_j\dfrac{\partial z^j}{\partial y^i}$, show that

$$u^i U_i = v^i V_i = w^i W_i$$

7. Show that $|A'^{ij}| = |A^{ij}|\left|\dfrac{\partial x}{\partial x'}\right|^{w-2}$, where A^{ij} and A_{ij} are the components of symmetric relative contravariant and covariant tensors, respectively, of weight w.

8. If a_{ij} are the components of a covariant tensor, show that the cofactors of the elements a_{ij} in $|a_{ij}|$ are the components of a relative contravariant tensor of weight 2.
9. Show that the Kronecker delta is a mixed tensor of rank two. [Hint: Use quotient law, taking an arbitrary vector.]
10. If $\dot{x}, \dot{y}\,; \ddot{x}, \ddot{y}$ are the components of velocity and acceleration, respectively, of a fluid element in Cartesian coordinates, find the corresponding quantities in plane polar coordinates.
11. Defining the conjugate component of a symmetric covariant tensor of second order, prove that the conjugate tensor fields a_{ij} and a^{ij} satisfy $a_{ij}\, a^{ij} = n$ in an n-dimensional Riemannian space.
12. Prove that the velocity of a fluid element is a contravariant vector of rank one, but acceleration is not a tensor in general. How do you define the acceleration so that it is a tensor?

3

Riemannian Metric and Fundamental Tensors

We have already introduced the concept of metric in Section 1.4 as the infinitesimal distance ds between two adjacent points in terms of a quadratic differential form.

3.1 Riemannian Metric

If x^i and $x^i + dx^i$ $(i = 1, 2, \ldots, n)$ are the two adjacent points of a coordinate system of n dimensions, then the infinitesimal distance ds between the points is defined by Riemann as

$$ds^2 = g_{ij}\, dx^i\, dx^j, \tag{3.1.1}$$

where g_{ij}'s are the functions of the coordinates x^i subject to the restriction that the determinant $g = |g_{ij}| \neq 0$. At this stage, g_{ij}'s are termed as metric functions. For real quadratic differential, it is assumed to be positive definite (as discussed in Section 1.3). But it needs to mention that the metric may not always be definite as in the case of general theory of relativity which may be indefinite also. The metric defined by Riemann above is an extension of classical metric to n dimensions and is called the Riemannian metric.

The geometry based on a Riemannian metric is called the **Riemannian geometry**, and the space whose geometry is developed on the basis of such a metric is called the **Riemannian space**.

3.2 Cartesian Coordinate System and Orthogonal Coordinate System

If the metric (3.1.1) is reducible to a particular coordinate system y^i by means of a transformation so that

$$ds^2 = a_{ij}dy^i dy^j \tag{3.2.1}$$

where a_{ij} are constants, the system is called a Cartesian coordinate system. On the other hand, if $a_{ij} = 0$ for $i \neq j$, the coordinate system is called **orthogonal coordinate system.**

3.3 Euclidean Space of *n* dimensions, Euclidean Co-ordinates, and Euclidean Geometry

If the metric (3.2.1) in terms of the coordinates y^i reduces in particular to the form

$$ds^2 = \sum_{i=1}^{n} \delta^i_j dy^i dy^j = \sum_{i=1}^{n} \left(dy^i\right)^2 \tag{3.3.1}$$

for $a_{ij} = \delta^i_j$, it is called the **Euclidean metric,** and the corresponding space is called the **Euclidean space.** The geometry developed on the basis of Euclidean metric is called the **Euclidean geometry.** Here, y^i's (when expressed in (3.3.1) form) are called the Euclidean coordinates.

Definition

Element of length: The element of length ds (in conformity with the concept of metric) is defined as $ds^2 = \in g_{ij} dx^i dx^j$, where $\in$ is ± 1 so as to make the right hand side positive. Therefore, the element of length "ds" is the magnitude of a contravariant vector, $dx^i \left[= \lambda^i (\text{say}) \right]$ so that $\lambda^2 = \in g_{ij} \lambda^i \lambda^j$ (will be seen later). It represents the magnitude of any contravariant vector characterized by the vector components λ^i.

Therefore, ds^2 from its nature being the intrinsic concept of a space and correlating with the magnitude of a vector must be **invariant.**

3.4 The Metric Functions g_{ij} Are Second-Order Covariant Symmetric Tensors

The metric $ds^2 = g_{ij} dx^i dx^j$ in Riemannian space is invariant.

So,

$$(ds^2 =) g'_{\alpha\beta} dx'^{\alpha}\, dx'^{\beta} = g_{ij} dx^i dx^j \quad \text{in } x'^{\alpha} \text{ and } x^i \text{ systems}$$

$$= g_{ij} \frac{\partial x^i}{\partial x'^{\alpha}} dx'^{\alpha} \frac{\partial x^j}{\partial x'^{\beta}} dx'^{\beta}$$

$$\left(g'_{\alpha\beta} - g_{ij} \frac{\partial x^i}{\partial x'^{\alpha}} \frac{\partial x^j}{\partial x'^{\beta}} \right) dx'^{\alpha} dx'^{\beta} = 0$$

$$g'_{\alpha\beta} = g_{ij} \frac{\partial x^i}{\partial x'^{\alpha}} \frac{\partial x^j}{\partial x'^{\beta}} \quad (\because dx'^{\alpha} \text{ is arbitrary}),$$

which is the transformation law of a covariant second-order tensor.

Hence, g_{ij} is a second-order covariant tensor.

Subject to the conformability of addition and subtraction of tensors, g_{ij} can be written as

$$g_{ij} = \frac{1}{2}\left(g_{ij} + g_{ji}\right) + \frac{1}{2}\left(g_{ij} - g_{ji}\right)$$

$$= A_{ij} + B_{ij},$$

putting $\frac{1}{2}\left(g_{ij} + g_{ji}\right) = A_{ij}$ and $\frac{1}{2}\left(g_{ij} - g_{ji}\right) = B_{ij}$.

Interchanging i and j in both A_{ij} and B_{ij}, we get

$$A_{ji} = \frac{1}{2}\left(g_{ji} + g_{ij}\right) = A_{ij}.$$

Hence, it is symmetric, and $B_{ji} = \frac{1}{2}\left(g_{ji} - g_{ij}\right) = -\frac{1}{2}\left(g_{ij} - g_{ji}\right) = -B_{ij}$; hence, B_{ij} is antisymmetric.

$$\therefore g_{ij} dx^i dx^j = \left(A_{ij} + B_{ij}\right) dx^i dx^j = A_{ij} dx^i dx^j + B_{ij} dx^i dx^j$$
$$= A_{ij} dx^i dx^j + 0. \tag{3.4.1}$$

For

$$B_{ij} dx^i dx^j = B_{ji} dx^j dx^i \left(i \leftrightarrow j\right)$$

$$= -B_{ij} dx^i dx^j.$$

$\because B_{ij}$ is antisymmetric

or $2B_{ij} dx^i dx^j = 0$

or $B_{ij} dx^i dx^j = 0$

Hence, from (3.4.1), $(g_{ij} - A_{ij})dx^i dx^j = 0.$

$$\therefore g_{ij} = A_{ij},$$

which is a symmetric tensor, $\because$ dx^i is arbitrary.

$\therefore$ The metric functions g_{ij} are the second-order symmetric covariant tensors. They are also called **fundamental covariant** tensors.

Note: The readers are advised to explore the use of the nomenclature "**fundamental**" for the tensor g_{ij}.

For the evolution of "tensor calculus," the covariant and contravariant tensors (yielding mixed tensor) of some kind it should go together hand in hand. So, it demands to define a contravariant tensor as the counterpart of g_{ij} to **supplement** the development of the branch.

Some applicable results:

1. If $g = |g_{ij}| \neq 0$ and G^{jk} are the cofactors of g_{kj} in g, then from the properties of determinant connecting cofactors, we have

$$g_{ij}G^{jk} = g \quad \text{when } i = k$$
$$= 0 \quad \text{when } i \neq k.$$

Otherwise, $g_{ij}G^{ik} = g\delta_j^k$.

2. If λ_{ij} is a second-order covariant tensor and u^j an arbitrary vector, then their inner product $\lambda_{ij}u^j$ is a covariant vector v_i, i.e., $\lambda_{ij}u^j = v_i$ but $g_{ij}u^j = u_i$.

   The difference is to be **noted down** for conception (!). In this case, u_i is called the **associate vector** to u^j by means of the fundamental tensor g_{ij}.

   Similarly, if g^{ij} is the second fundamental tensor (it will be proved in Section 3.5) and u_j (associate to u^i by means of g_{ij}) is an arbitrary vector, then $g^{ij}u_j = u^i$, a contravariant vector. Thus,

3. The **inner multiplication** of any tensors (vectors) contravariant and covariant by g_{ij} and g^{ij}, respectively, is to be treated as a means to **lower down** or **raise the indices** in tensors.

Definition

The set of n^2 functions g^{ij} is defined as

$$g^{ij} = \frac{\text{cofactor of } g_{ij} \text{ in the determinant } g}{g},$$

where

$$g = |g_{ij}| \neq 0$$

$$= \frac{G^{ji}}{g}, G^{ji},$$

written conventionally to denote the cofactor of g_{ij} in g.

It can be proved that the class of functions g^{ij} is a second-order symmetric contravariant tensor.

3.5 The Function g^{ij} Is a Contravariant Second-Order Symmetric Tensor

Let us consider the covariant vector (tensor) u_j by means of its associate vector u^k so that $u_j = g_{kj} u^k$, where u^k is an arbitrary vector.

Now,

$$u_j g^{ij} = u_j \frac{G^{ji}}{g} = g_{kj} u^k \frac{G^{ji}}{g} = \left(g_{kj} G^{ji}\right) \frac{u^k}{g}$$

$$= g\delta_k^i \frac{u^k}{g} = u^i,$$

which is a vector (i.e., a tensor rank one).

∴ The inner product of the function g^{ij} with the arbitrary vector u_j is a tensor.

Hence, by **quotient law** of tensors, g^{ij} is a contravariant tensor of rank (order) two. It is also symmetric.

For g_{ij} is symmetric, $\Rightarrow G^{ji}$ is also symmetric.

$\Rightarrow \frac{G^{ji}}{g}$ is also symmetric, since the value of $g = |g_{ij}| \neq 0$ will not change the symmetric character on division except the numerical value.

∴ g^{ij} is also a second-order symmetric contravariant tensor, and it is also called "**fundamental tensor.**"

N.B.: In conformity with g_{ij}, g^{ij} a mixed-order tensor g_j^i is defined as $g_j^i = \delta_j^i$, it is also a tensor because δ_j^i is a mixed second-order tensor.

Example 1

Show that

i. $\left(g_{hj} g_{ik} - g_{hk} g_{ij}\right) g^{hj} = (n-1) g_{ik}$

ii. $\frac{\partial K}{\partial x^j}\left(g_{hk}g_{il} - g_{hl}g_{ik}\right)g^{hj} = \frac{\partial K}{\partial x^k}g_{il} - \frac{\partial K}{\partial x^l}g_{ik}$

i. $\text{L.H.S.} = \left(ng_{ik} - g_{hk}\delta_i^h\right) \quad \left(\because g_{hj}g^{hj} = \delta_h^h = n\right)$

$$= ng_{ik} - g_{ik}$$

$$= (n-1)g_{ik} = \text{R.H.S.}$$

Hence, proved.

ii. $\text{L.H.S.} = \frac{\partial K}{\partial x^j}\left(g_{hk}g_{il} - g_{hl}g_{ik}\right)g^{hj}$

$$= \frac{\partial K}{\partial x^j}\left(g_{hk}g^{hj}\right)g_{il} - \frac{\partial K}{\partial x^j}\left(g_{hl}g^{hj}\right)g_{ik}$$

$$= \frac{\partial K}{\partial x^j}\delta_k^j g_{il} - \frac{\partial K}{\partial x^j}\delta_l^j g_{ik}$$

$$= \frac{\partial K}{\partial x^k}g_{il} - \frac{\partial K}{\partial x^l}g_{ik} = \text{R.H.S.}$$

Hence, proved.

Example 2

Calculate g_{ij} for the coordinates x^i, given that

$$y^1 = x^1x^2\cos x^3$$

$$y^2 = x^1x^2\sin x^3$$

$$y^3 = \frac{1}{2}\left[\left(x^1\right)^2 + \left(x^2\right)^2\right]$$

where $x^1 \geq 0, x^2 \geq 0, 0 \leq x^3 \leq 2\pi$.

y^i is a Cartesian orthogonal coordinate system.

For a Cartesian coordinate system y^i, the metric is of the form $ds^2 = g'_{\alpha\beta}dy'^{\alpha}dy'^{\beta}$, where $g'_{\alpha\beta}$ are constants (for convenience of tensor, y' – dash is used in the metric).

But if it is orthogonal too, then

$$g'_{\alpha\beta} = 1 \text{ if } \alpha = \beta$$

$$= 0 \text{ if } \alpha \neq \beta.$$

$\therefore$ It can be expressed in the form:

$$ds^2 = g'_{ii}\left(dy'^i\right)^2, \quad g'_{\alpha\beta}=1 \text{ if } \alpha = \beta.$$

To make use of tensor notation, the given Cartesian orthogonal coordinate system y^i is replaced by x'^i so that $x'^1 = y^1$, $x'^2 = y^2$, $x'^3 = y^3$

$$g_{ij} = \frac{\partial x'^{\alpha}}{\partial x^{i}} \frac{\partial x'^{\beta}}{\partial x^{j}} g'_{\alpha\beta}$$

$$g_{ij} = \frac{\partial x'^{\alpha}}{\partial x^{i}} \frac{\partial x'^{\alpha}}{\partial x^{j}} g'_{\alpha\alpha}$$

$$\because g'_{\alpha\beta} = 1$$

exists when $\alpha = \beta$.

$$\therefore g_{11} = \frac{\partial x'^{1}}{\partial x^{1}} \frac{\partial x'^{1}}{\partial x^{1}} g'_{11} + \frac{\partial x'^{2}}{\partial x^{1}} \frac{\partial x'^{2}}{\partial x^{1}} g'_{22} + \frac{\partial x'^{3}}{\partial x^{1}} \frac{\partial x'^{3}}{\partial x^{1}} g'_{33}$$

$$= \left(x^{2} \cos x^{3}\right)^{2} 1 + \left(x^{2} \sin x^{3}\right)^{2} 1 + \left(x^{1}\right)^{2} \cdot 1$$

$$\left(\because y^{i} = x'^{i}\right)$$

$$= \left(x^{2}\right)^{2} + \left(x^{1}\right)^{2}$$

$$g_{22} = \frac{\partial x'^{1}}{\partial x^{2}} \frac{\partial x'^{1}}{\partial x^{2}} g'_{11} + \frac{\partial x'^{2}}{\partial x^{2}} \frac{\partial x'^{2}}{\partial x^{2}} g'_{22} + \frac{\partial x'^{3}}{\partial x^{2}} \frac{\partial x'^{3}}{\partial x^{2}} g'_{33}$$

$$= \left(x^{1} \cos x^{3}\right)^{2} \cdot 1 + \left(x^{1} \sin x^{3}\right)^{2} \cdot 1 + \left(x^{2}\right)^{2} \cdot 1$$

$$= \left(x^{1}\right)^{2} + \left(x^{2}\right)^{2}$$

$$g_{33} = \frac{\partial x'^{1}}{\partial x^{3}} \frac{\partial x'^{1}}{\partial x^{3}} g'_{11} + \frac{\partial x'^{2}}{\partial x^{3}} \frac{\partial x'^{2}}{\partial x^{3}} g'_{22} + \frac{\partial x'^{3}}{\partial x^{3}} \frac{\partial x'^{3}}{\partial x^{3}} g'_{33}$$

$$= \left(-x^{1} x^{2} \sin x^{3}\right)^{2} 1 + \left(x^{1} x^{2} \cos x^{3}\right)^{2} \cdot 1 + 0$$

$$= \left(x^{1} x^{2}\right)^{2}.$$

But

$$g_{13} = \frac{\partial x'^{1}}{\partial x^{1}} \frac{\partial x'^{1}}{\partial x^{3}} g'_{11} + \frac{\partial x'^{2}}{\partial x^{1}} \frac{\partial x'^{2}}{\partial x^{3}} g'_{22} + \frac{\partial x'^{3}}{\partial x^{1}} \frac{\partial x'^{3}}{\partial x^{3}} g'_{33}$$

$$= \left(x^{2} \cos x^{3}\right)\left(-x^{1} x^{2} \sin x^{3}\right) + \left(x^{2} \sin x^{3}\right)\left(x^{1} x^{2} \cos x^{3}\right) \cdot 1 + x^{1} \times 0$$

$$= -x^{1}\left(x^{2}\right)^{2} \cos x^{3} \sin x^{3} + x^{1}\left(x^{2}\right)^{2} \cos x^{3} \sin x^{3}$$

$$= 0$$

$$\because 0 \le x^{3} \le 2\pi \text{ and } x^{1} \ge 0$$

and

$$g_{23} = \frac{\partial x'^1}{\partial x^2}\frac{\partial x'^1}{\partial x^3} g'_{11} + \frac{\partial x'^2}{\partial x^2}\frac{\partial x'^2}{\partial x^3} g'_{22} + \frac{\partial x'^3}{\partial x^2}\frac{\partial x'^3}{\partial x^3} g'_{33}$$

$$= x^1 \cos x^3 \left(-x^1 x^2 \sin x^3\right)\cdot 1 + x^1 \sin x^3 \left(x^1 x^2 \cos x^3\right) + 0$$

$$= -\left(x^1\right)^2 x^2 \cos x^3 \sin x^3 + \left(x^1\right)^2 x^2 \cos x^3 \sin x^3$$

$$\because 0 \le x^3 \le 2\pi$$

$$= 0.$$

Thus, $g_{11} = g_{22} = \left(x^1\right)^2 + \left(x^2\right)^2, g_{12} = 2x^1x^2, g_{33} = \left(x^1x^2\right)^2$, and $g_{13} = g_{23} = 0$.

Example 3

If the metric in V_2 is such that $ds^2 = Edu^2 + 2Fdudv + Gdv^2$, then find the values of $g^{ij}\,(i, j = 1, 2)$.

The given metric in V_2 is $ds^2 = Edu^2 + Fdudv + Fdvdu + Gdv^2$

$$\therefore |g_{ij}| = g = \begin{vmatrix} E & F \\ F & G \end{vmatrix} = EG - F^2 \ne 0,$$

where

$$g_{11} = E, g_{12} = g_{21} = F, \text{ and } g_{22} = G.$$

But

$$g^{ij} = \frac{\text{cofactor of } g_{ij} \text{ in } g}{g} \quad (g \ne 0)$$

$$\therefore g^{11} = \frac{\text{cofactor of } g_{11} \text{ in } g}{g} = \frac{G}{g}$$

$$g^{12} = g^{21} = \frac{\text{cofactor of } g_{12} \text{ or } g_{21} \text{ in } g}{g}$$

$$= \frac{-F}{g}$$

and

$$g^{22} = \frac{\text{cofactor of } g_{22} \text{ in } g}{g} = \frac{E}{g}$$

$$\therefore g^{11} = \frac{G}{g}, \quad g^{12} = g^{21} = \frac{-F}{g}$$

and

$$g^{22} = \frac{E}{g},$$

$$g = EG - F^2 \ne 0.$$

Example 4

Determine the conjugate metric tensor of g_{ij} in the spherical coordinate system.

In the spherical coordinate system, the metric is given by

$$ds^2 = dr^2 + r^2 d\theta^2 + r^2 \sin^2 \theta d\phi^2.$$

Comparing it with $ds^2 = g_{ij}dx^i dx^j = g_{11}dx^1dx^1 + g_{22}dx^2dx^2 + g_{33}dx^3dx^3$, we can write

$$ds^2 = \left(dx^1\right)^2 + r^2\left(dx^2\right)^2 + r^2 \sin^2 \theta \left(dx^3\right)^2$$

in the given metric, where $x^1 = r$, $x^2 = \theta$, $x^3 = \phi$.

$$\therefore g_{11} = 1,\ g_{22} = r^2,\ g_{33} = r^2 \sin^2 \theta, \text{ and } g_{12} = g_{23} = g_{13} = 0$$

$$\therefore g = \begin{vmatrix} 1 & 0 & 0 \\ 0 & r^2 & 0 \\ 0 & 0 & r^2 \sin^2 \theta \end{vmatrix} = r^4 \sin^2 \theta \neq 0.$$

Now, by definition,

$$g^{11} = \frac{\text{cofactor of } g_{11} \text{ in } g}{g} = \frac{1}{r^4 \sin^2 \theta} \times \begin{vmatrix} r^2 & 0 \\ 0 & r^2 \sin^2 \theta \end{vmatrix}$$

$$= \frac{r^4 \sin^2 \theta}{r^4 \sin^2 \theta} = 1$$

$$g^{22} = \frac{\text{cofactor of } g_{22} \text{ in } g}{g} = \frac{1}{r^4 \sin^2 \theta} \times \begin{vmatrix} 1 & 0 \\ 0 & r^2 \sin^2 \theta \end{vmatrix}$$

$$= \frac{r^2 \sin^2 \theta}{r^4 \sin^2 \theta} = \frac{1}{r^2}$$

$$g^{33} = \frac{\text{cofactor of } g_{33} \text{ in } g}{g} = \frac{1}{r^4 \sin^2 \theta} \times \begin{vmatrix} 1 & 0 \\ 0 & r^2 \end{vmatrix}$$

$$= \frac{r^2}{r^4 \sin^2 \theta} = \frac{1}{r^2 \sin^2 \theta}$$

and $g^{ij} = 0$ for $i \neq j$.

3.6 Scalar Product and Magnitude of Vectors

If $\vec{a}(a^i)$ and $\vec{b}(b^i)$ are the two non-null vectors, the scalar product of the two vectors is defined as

$$\begin{aligned}\bar{a}\cdot\bar{b} &= g_{ij}a^i b^j \\ &= a^i b_i = a_j b^j\end{aligned} \tag{3.6.1}$$

in terms of associate vectors of a^i and b^j.

If a_i and b_i are their covariant components, the scalar product of $\bar{a}$ and $\bar{b}$ can also be defined as

$$\bar{a}\cdot\bar{b} = g^{ij}a_i b_j \left(a^j b_j \text{ or } a_j b^j\right).$$

On the other hand, if the scalar product of (say) $\vec{a}$ is considered in itself, then

$$\bar{a}\cdot\bar{a} = g_{ij}a^i a^j = a_j a^j = a^2 \tag{3.6.2}$$

which is the square of the magnitude of the vector $\vec{a}$.

If we recall the metric in a V_n, namely, $ds^2 = g_{ij}dx^i dx^j$ and $t^i = \dfrac{dx^i}{ds}$ are the components of a unit tangent at a point P to a curve C of V_n, the metric can be simplified to give

$$\begin{aligned}1 &= g_{ij}\frac{dx^i}{ds}\frac{dx^j}{ds} \\ &= g_{ij}t^i t^j,\end{aligned}$$

which confirms the unit magnitude (defined above) of t^i.

N.B.: If for a curve or for a portion of a curve $g_{ij}\dfrac{dx^i}{ds}\dfrac{dx^j}{ds} = 0$ that is of length (magnitude) zero or minimal, the lines of zero length are identified as the **world lines of light** in space–time continuum of general relativity.

3.7 Angle between Two Vectors and Orthogonal Condition

If a and b are the magnitudes of the two non-null and nonparallel vectors $\vec{a}$ and $\vec{b}$, respectively, with inclination θ, then $\dfrac{\vec{a}}{a}$ and $\dfrac{\vec{b}}{b}$ are the unit vectors to represent their directions.

The cosine of the angle between these two unit vectors (in analogy with Euclidean space of three dimensions E_3) in a Riemannian V_n is defined as

$$\cos\theta = \frac{\vec{a}}{a}\cdot\frac{\vec{b}}{b} = \frac{\bar{a}\cdot\bar{b}}{ab} = \frac{g_{ij}a^i b^j}{\sqrt{g_{ij}a^i a^j}\sqrt{g_{ij}b^i b^j}} \tag{3.7.1}$$

using (3.6.1) and (3.6.2).

For orthogonality of the vectors $\cos\theta = 0$ which gives the mathematical expression $g_{ij}a^i b^j = 0$; otherwise, $a_j b^j = 0$ when the two vectors are orthogonal.

If $\hat{a}$ and $\hat{b}$ are the unit vectors so that

$$\sqrt{g_{ij}a^i a^j} = 1 = \sqrt{g_{ij}b^i b^j}, \text{then } \cos\theta = g_{ij}a^i b^j = a_j b^j.$$

But for a real positive definite fundamental form of a Riemannian space V_n, the numerical value of $\cos\theta$ **must not be greater than unity**. This can be proved as follows:

The square of the vector $m\hat{a} + n\hat{b}$ determined by the pencils a^i and b^i can be written as

$$\begin{aligned}
\lambda^2 &= g_{ij}(ma^i + nb^i)(ma^j + nb^j) \quad [= g_{ij}\lambda^i\lambda^j] \\
&= g_{ij}a^i a^j m^2 + g_{ij}a^i b^j mn + \underset{(i\leftrightarrow j)}{g_{ij}\, b^i a^j}\, mn + n^2 g_{ij}b^i b^j \\
&= m^2 + 2g_{ij}a^i b^j mn + n^2 \\
&= m^2\left[1 + 2\frac{n}{m}g_{ij}a^i b^j + \left(\frac{n}{m}\right)^2\right] \\
&= m^2\left[\left(\frac{n}{m} + g_{ij}a^i b^j\right)^2 + 1 - \left(g_{ij}a^i b^j\right)^2\right].
\end{aligned}$$

Since for all $\frac{n}{m}$, this value should be positive, so $\left(g_{ij}a^i b^j\right)^2 \leq 1$.

$$\therefore \cos^2\theta \leq 1.$$

$\therefore$ The angle θ defined above is real for a positive definite fundamental form of Riemannian V_n.

Exercises

1. If $\bar{u}(u_i)$ and $\bar{v}(v_i)$ are the two vectors, then show that the quantities $A_{ij} = u_i v_j - u_j v_i$ are the components of a skew-symmetric covariant

tensor of order two. What is the connection of it with the cross product of the two vectors? [Hint: Interchanges i and j.]

2. Prove that $\sqrt{g}dx^1dx^2\ldots dx^n = \sqrt{g'}dx'^1dx'^2\ldots dx'^n$ for an element of volume in V_n.
3. Determine the components of the fundamental contravariant tensor g^{ij} for
   i. $ds^2 = dr^2 + r^2d\theta^2 + dz^2$
   ii. $ds^2 = dr^2 + r^2d\theta^2 + r^2\sin^2\theta d\phi^2$.
4. Prove that the cosine of angle between two coordinate parametric curves x^i = constant and x^j = constant is $\frac{g^{ij}}{\sqrt{g^{ii}}\sqrt{g^{jj}}}$. [Hint: Angle between the curves is the angle between the normal vectors ∇x^i, ∇x^j.]
5. Show that the angle between two contravariant vectors is real when the Riemannian metric is positive definite.

4

Christoffel Three-Index Symbols (Brackets) and Covariant Differentiation

Notations and symbols are some fundamental ingredients for strong and beautiful representation of mathematical concepts. Tensor is the embodiment of such shorthand symbols and notations, which can concisely and forcefully uphold mathematical expressions as described in this chapter.

4.1 Christoffel Symbols (or Brackets) of the First and Second Kinds

If g_{ij} and g^{ij} be the first and second kinds of fundamental tensors, respectively, then the Christoffel symbols (or brackets) of the first and second kinds are, respectively, defined as

$$[ij,k] = \Gamma_{ij,k} = \tfrac{1}{2}\left[\frac{\partial g_{ik}}{\partial x^j} + \frac{\partial g_{jk}}{\partial x^i} - \frac{\partial g_{ij}}{\partial x^k}\right] \tag{4.1.1}$$

and

$$\left\{ {i \atop jk} \right\} = \Gamma^i_{jk} = g^{i\alpha}\Gamma_{jk,\alpha} \tag{4.1.2}$$

Since g_{ij} is symmetric, $\Gamma_{ij,k}$ is also symmetric, which is obvious from its nature of occurrence in (4.1.1) with respect to i and j indices. Also, as $g^{i\alpha}$ and $\Gamma_{jk,\alpha}$ are both symmetric, Γ^i_{jk} is also symmetric with respect to j and k indices.

Observation: From the nature of definition of Γ^i_{ik} in (4.1.2), we can promptly verify that $g_{i\beta}\Gamma^i_{jk} = g_{i\beta}\, g^{i\alpha}\Gamma_{jk,\alpha} = \delta^\alpha_\beta \Gamma_{jk,\alpha} = \Gamma_{jk,\beta}$ which is the first kind.

N.B.: The author will make use of the symbols $\Gamma_{ij,k}$ and Γ^i_{jk} throughout this book instead of the following old symbols: $[ij,k]$ and $\left\{ {i \atop jk} \right\}$.

4.2 Two Standard Applicable Results of Christoffel Symbols

i. $\Gamma_{ij,k} + \Gamma_{jk,i} = \frac{\partial g_{ik}}{\partial x^j}$.

ii. $\Gamma_{ij}^{i} = \frac{\partial}{\partial x^j}\left(\log\sqrt{\pm g}\right)$ $(g > 0, g < 0)$.

Proof of (i)

$$\Gamma_{ij,k} + \Gamma_{jk,i} = \frac{1}{2}\left[\frac{\partial g_{ik}}{\partial x^j} + \frac{\partial g_{jk}}{\partial x^i} - \frac{\partial g_{ij}}{\partial x^k}\right] + \frac{1}{2}\left[\frac{\partial g_{ji}}{\partial x^k} + \frac{\partial g_{ki}}{\partial x^j} - \frac{\partial g_{jk}}{\partial x^i}\right]$$

i. $$= \frac{1}{2} \times 2\frac{\partial g_{ik}}{\partial x^j},\ g_{ij} = g_{ji}, \text{etc.}$$

$$= \frac{\partial g_{ik}}{\partial x^j}.$$

Hence, proved.

Proof of (ii)

Differentiating partially the elements of the determinant $g = |g_{ij}|$ row wise with respect to x^j, and expanding each of the n determinants in terms of the elements of that particular differentiated row, it can be precisely written as a whole:

$$\frac{\partial g}{\partial x^j} = \frac{\partial g_{ik}}{\partial x^j} G^{ki},$$

where G^{ki} are the cofactors of g_{ik} in g ($i, k = 1, 2,\dots, n$).

$$= \frac{\partial g_{ik}}{\partial x^j}\ g\ g^{ik}$$

$$\frac{1}{g}\frac{\partial g}{\partial x^j} = g^{ik}\left[\Gamma_{ij,k} + \Gamma_{jk,i}\right], (g \neq 0)$$

using the result (i) above.

$$= g^{ik}\Gamma_{ij,k} + g^{ki}\Gamma_{jk,i} \quad g^{ik} = g^{ki}$$

$$= \Gamma_{ij}^{i} + \underset{(k\to i)}{\Gamma_{jk}^{k}} = 2\Gamma_{ij}^{i}$$

$$\Gamma^{i}_{ij} = \frac{1}{2g}\frac{\partial g}{\partial x^j} = \frac{\partial}{\partial x^j}\left(\log\sqrt{g}\right) \quad \text{if } g > 0$$

$$= \frac{\partial}{\partial x^j}\left(\log\sqrt{-g}\right) \quad \text{if } g < 0.$$

Hence, proved.

4.3 Evolutionary Basis of Christoffel Symbols (Brackets)

In the **Cartesian coordinate** system, the law of parallel displacement takes the form:

$$a_{i,k}\delta x^k = 0$$

$$\text{i.e.,} \quad \frac{\partial a^i}{\partial x^k}\delta x^k = 0 \quad \because \Gamma^{\alpha}_{ik} = 0' \tag{4.3.1}$$

where δx^k represents the infinitesimal displacement. Let us transform the vector components a_i to a new coordinate system (ξ^i). Then, we have

$$a_i = \frac{\partial \xi^r}{\partial x^i}a'_r$$

$$\therefore \frac{\partial a_i}{\partial x^k} = \frac{\partial}{\partial \xi^s}\left(\frac{\partial \xi^r}{\partial x^i}a'_r\right)\frac{\partial \xi^s}{\partial x^k}$$

$$= \frac{\partial \xi^s}{\partial x^k}\frac{\partial \xi^r}{\partial x^i}\frac{\partial a'_r}{\partial \xi^s} + \frac{\partial \xi^s}{\partial x^k}\frac{\partial x^l}{\partial \xi^s}\frac{\partial^2 \xi^r}{\partial x^l \partial x^i}a'_r \tag{4.3.2}$$

$$\therefore 0 = \frac{\partial a_i}{\partial x^k}\delta x^k = \left(\frac{\partial \xi^r}{\partial x^i}\frac{\partial a'_r}{\partial \xi^s} + \frac{\partial x^l}{\partial \xi^s}\frac{\partial^2 \xi^r}{\partial x^l \partial x^i}a'_r\right)\frac{\partial \xi^s}{\partial x^k}\delta x^k$$

$$\therefore 0 = a_{i,k}\delta x^k = \left(\frac{\partial \xi^r}{\partial x^i}\frac{\partial a'_r}{\partial \xi^s} + \frac{\partial x^l}{\partial \xi^s}\frac{\partial^2 \xi^r}{\partial x^l \partial x^i}a'_r\right)\delta \xi^s$$

$$\left(\because \delta \xi^s = \frac{\partial \xi^s}{\partial x^k}\delta x^k\right).$$

$(\delta a'_r =)\dfrac{\partial a'_r}{\partial \xi^s}\delta \xi^s$ is the actual increment of a'_r as a result of the displacement, and it is denoted by $\delta a'_r$.

Multiplying the R.H.S. of (4.3.2) by $\frac{\partial x^i}{\partial \xi^{l}}$, we get

$$\delta_l^r \delta a_r' = -\frac{\partial x^l}{\partial \xi^s}\frac{\partial x^i}{\partial \xi^l}\frac{\partial^2 \xi^r}{\partial x^l \partial x^i} a_r' \delta \xi^s$$

$$\Rightarrow \delta a_l' = -\frac{\partial x^l}{\partial \xi^s}\frac{\partial x^i}{\partial \xi^l}\frac{\partial^2 \xi^r}{\partial x^i \partial x^l} a_r' \delta \xi^s. \tag{4.3.3}$$

When no Cartesian coordinate system can be introduced, we shall retain the linear form of Equation (4.3.3) and assume that, because of a parallel displacement, the infinitesimal changes of the vector components are bilinear functions of the vector components, and the components of the infinitesimal displacement are **defined** by

$$\delta a^i = -\overset{I}{\Gamma}{}^i_{kl} a^k \delta \xi^l. \tag{4.3.4}$$

$$\delta a_k = +\overset{II}{\Gamma}{}^i_{kl} a_i \delta \xi^l. \tag{4.3.5}$$

The coefficients $\overset{I}{\Gamma}{}^i_{kl}$ and $\overset{II}{\Gamma}{}^i_{kl}$ of these new tentative laws are, so far, entirely unknown quantities. But we can determine their transformation laws, $\delta a'^k$ as the difference between two vectors at two points, characterized by the coordinate values ξ^1 and $\xi^1 + \delta\xi^1$.

In case of a coordinate transformation, the new $\delta a'^k$ is given by

$$\begin{aligned} \delta a'^k &= \left(\frac{\partial \xi'^k}{\partial \xi^s} a^s\right)_{\xi^l+\delta\xi^l} - \left(\frac{\partial \xi'^k}{\partial \xi^s} a^s\right)_{\xi^l} \left(a'^k = \frac{\partial \xi'^k}{\partial \xi^s} a^s\right) \\ &= \frac{\partial}{\partial \xi^l}\left(\frac{\partial \xi'^k}{\partial \xi^s} a^s\right)\delta\xi^l \\ &= \frac{\partial^2 \xi'^k}{\partial \xi^l \partial \xi^s} a^s \delta \xi^l + \frac{\partial \xi'^k}{\partial \xi^s} a^s_{,l} \delta \xi^l \quad a^s_{,l} = \frac{\partial a^s}{\partial \xi^l} \\ \delta a'^k &= \frac{\partial^2 \xi'^k}{\partial \xi^l \partial \xi^s} a^s \delta \xi^l + \frac{\partial \xi'^k}{\partial \xi^s} \delta a^s. \end{aligned} \tag{4.3.6}$$

Using (4.3.6) into the L.H.S. of the following equation corresponding to (4.3.4),

$$\delta a'^k = \overset{I}{\Gamma}{}'^k_{mn} a'^m \delta \xi'^n$$

and replacing a'^m and $\delta\xi'^n$ by the expressions
$a'^m = \frac{\delta\xi'^m}{\partial\xi^s} a^s$ and $\delta\xi'^n = \frac{\delta\xi'^n}{\delta\xi^l}\delta\xi^l$, we get

$$\frac{\partial^2 \xi'^k}{\partial\xi^l \partial\xi^s} a^s \delta\xi^l + \frac{\partial\xi'^k}{\partial\xi^s}\delta a^s = -\overset{I}{\underset{mn}{\Gamma'^k}} \frac{\delta\xi'^m}{\partial\xi^s} a^s \frac{\delta\xi'^n}{\delta\xi^l}\delta\xi^l.$$

Substituting the value of δa^s from Equation (4.3.4) into it, we get

$$\frac{\partial^2 \xi'^k}{\partial\xi^l \partial\xi^s} a^s \delta\xi^l - \underset{(s\to r)}{\frac{\partial\xi'^k}{\partial\xi^s}} \underset{(m\to s)(n\to l)}{\overset{I}{\Gamma^s_{mn}} a^m \delta\xi^n} = -\overset{I}{\underset{mn}{\Gamma'^k}} \frac{\partial\xi'^m}{\partial\xi^s} a^s \frac{\partial\xi'^n}{\partial\xi^l}\delta\xi^l$$

$$\therefore \left(\frac{\partial^2 \xi'^k}{\partial\xi^l \partial\xi^s} - \frac{\partial\xi'^k}{\partial\xi^r} \overset{I}{\underset{sl}{\Gamma^r}} \right) a^s \delta\xi^l = -\overset{I}{\underset{mn}{\Gamma'^k}} \frac{\partial\xi'^m}{\partial\xi^s} \frac{\partial\xi'^n}{\partial\xi^l} a^s \delta\xi^l.$$

Hence, $\frac{\partial\xi'^m}{\partial\xi^s}\frac{\partial\xi'^n}{\partial\xi^l}\overset{I}{\underset{mn}{\Gamma'^k}} = \frac{\partial\xi'^k}{\partial\xi^r}\overset{I}{\underset{sl}{\Gamma^r}} - \frac{\partial^2 \xi'^k}{\partial\xi^l \partial\xi^s}$, since a^s and $\delta\xi^l$ are arbitrary.

Multiplying by $\frac{\partial\xi^s}{\partial\xi'^a}\frac{\partial\xi^l}{\partial\xi'^b}$, we get the transformation formula $\overset{I}{\underset{sl}{\Gamma^r}}$ as

$$\overset{I}{\underset{ab}{\Gamma'^k}} = \frac{\partial\xi^s}{\partial\xi'^a}\frac{\partial\xi^l}{\partial\xi'^b}\left(\frac{\partial\xi'^k}{\partial\xi^r}\overset{I}{\underset{sl}{\Gamma^r}} - \frac{\partial^2 \xi'^k}{\partial\xi^l \partial\xi^s} \right). \tag{4.3.7}$$

The last term of the R.H.S. of (4.3.7) can be written as

$$-\frac{\partial\xi^s}{\partial\xi'^a}\frac{\partial\xi^l}{\partial\xi'^b}\frac{\partial^2 \xi'^k}{\partial\xi^l \partial\xi^s} = -\frac{\partial\xi^s}{\partial\xi'^a}\frac{\partial}{\partial\xi^s}\left(\frac{\partial\xi^l}{\partial\xi'^b}\frac{\partial\xi'^k}{\partial\xi^l} \right) + \frac{\partial\xi^s}{\partial\xi'^a}\frac{\partial\xi'^k}{\delta\xi^l}\frac{\partial}{\partial\xi^s}\left(\frac{\partial\xi^l}{\partial\xi'^b} \right)$$

$$= \frac{\partial\xi^s}{\partial\xi'^a}\frac{\partial}{\partial\xi^s}\left(\delta_b'^k\right) + \frac{\partial\xi'^k}{\partial\xi^l}\frac{\partial^2 \xi^l}{\partial\xi'^a \partial\xi'^b}$$

$$= 0 + \frac{\partial\xi'^k}{\partial\xi^l}\frac{\partial^2 \xi^l}{\partial\xi'^a \partial\xi'^b}.$$

($\because \delta_b^a$ is constant)

$$\therefore -\frac{\partial\xi^s}{\partial\xi'^a}\frac{\partial\xi^1}{\partial\xi'^b}\frac{\partial^2 \xi'^k}{\partial\xi^l \partial\xi^s} = \frac{\partial\xi'^k}{\partial\xi^l}\frac{\partial^2 \xi^1}{\partial\xi'^a \partial\xi'^b}.$$

Therefore, Equation (4.3.7) becomes

$$\overset{I}{\Gamma}{}'^{k}_{ab} = \underset{(s\to r)}{\frac{\partial \xi^{s}}{\partial \xi'^{a}}}\underset{(l\to s)}{\frac{\partial \xi^{l}}{\partial \xi'^{b}}}\underset{(r\to l)}{\frac{\partial \xi'^{k}}{\partial \xi^{r}}}\overset{I}{\Gamma}{}^{r}_{sl} + \frac{\partial \xi'^{k}}{\partial \xi^{l}}\frac{\partial^{2} \xi^{l}}{\partial \xi'^{a}\partial \xi'^{b}}$$

$$\therefore \overset{I}{\Gamma}{}'^{k}_{ab} = \frac{\partial \xi'^{k}}{\partial \xi^{l}}\left(\frac{\partial \xi^{r}}{\partial \xi'^{a}}\frac{\partial \xi^{s}}{\partial \xi'^{b}}\overset{I}{\Gamma}{}^{l}_{rs} + \frac{\partial^{2} \xi^{l}}{\partial \xi'^{a}\partial \xi'^{b}}\right). \tag{4.3.8}$$

Similarly, we can obtain the transformation law for $\overset{II}{\Gamma}{}^{k}_{ab}$. It is identical with $\overset{I}{\Gamma}{}^{k}_{ab}$. The transformation law consists of two terms. The first term depends on the $\overset{I}{\Gamma}{}^{k}_{ab}$ in the old coordinate system, and the second term does not depend on $\overset{I}{\Gamma}{}^{k}_{ab}$ and adds an expression which is symmetric in two subscripts. So, even though the $\overset{I}{\Gamma}{}^{k}_{ab}$ may vanish in one coordinate system, they do not vanish in other systems. But if the $\overset{I}{\Gamma}{}^{k}_{ab}$ were **symmetric** in their subscripts in one coordinate system, they would be symmetric in every other coordinate system as well.

This would be particularly true if the $\overset{I}{\Gamma}{}^{k}_{ab}$ were to vanish in one system. Besides, if $\overset{I}{\Gamma}{}^{k}_{ab}$ were equal to the $\overset{II}{\Gamma}{}^{k}_{ab}$ in one coordinate system, this **equality would be preserved** by arbitrary coordinate transformations. We shall find that geometrical considerations of systems in which $\overset{I}{\Gamma}{}^{k}_{ab}$ satisfies both these conditions.

Let us displace two vectors a_i and b^i parallel to themselves along an infinitesimal path $\delta\xi^{i}$. The change of their scalar product, $a_i b^i$, is given by

$$\begin{aligned}\delta(a_i b^i) &= a_i \delta b^i + b^i \delta a_i \\ &= a_i\left(-\overset{I}{\Gamma}{}^{i}_{kl} b^k \delta\xi^{l}\right) + b^i \underset{(i\leftrightarrow k)}{\overset{II}{\Gamma}{}^{k}_{il}} a_k \delta\xi^{l} \\ &= -\overset{I}{\Gamma}{}^{i}_{kl} a_i b^k \delta\xi^{l} + \overset{II}{\Gamma}{}^{i}_{kl} a_i b^k \delta\xi^{l} \\ &= a_i b^k\left(\overset{II}{\Gamma}{}^{i}_{kl} - \overset{I}{\Gamma}{}^{i}_{kl}\right)\delta\xi^{l}.\end{aligned} \tag{4.3.9a}$$

When two vectors are displaced parallel to themselves, their scalar product always **remains constant** if and only if $\overset{I}{\Gamma}{}^{i}_{kl}$ are equal to $\overset{II}{\Gamma}{}^{i}_{kl}$.

If the law of parallel displacement of Equations (4.3.4) and (4.3.5) is extended from vectors to tensors, we can displace any tensor parallel to itself according to the rule:

$$t_{ik}^{l} = a^{l} b_{i} c_{k} \left(\text{tensor of rank } 3\right). \tag{4.3.9b}$$

$$\delta t_{ik}^{l} = \delta a^{l} b_{i} c_{k} + a^{l} \delta b_{i} c_{k} + a^{l} b_{i} \delta c_{k}$$

$$= -\overset{I}{\underset{rs}{\Gamma^{l}}} a^{r} b_{i} c_{k} \delta\xi^{s} + \overset{II}{\underset{is}{\Gamma^{r}}} b_{r} a^{l} c_{k} \delta\xi^{s} + \overset{II}{\underset{ks}{\Gamma^{r}}} c_{r} a^{l} b_{i} \delta\xi^{s}$$

$$= \left(\overset{II}{\Gamma^{r}_{is}} t^{l}_{rk} + \overset{II}{\Gamma^{r}_{ks}} t^{l}_{ri} - \overset{I}{\Gamma^{l}_{rs}} t^{r}_{ik} \right) \delta\xi^{s},$$

making use of (4.3.9b)

$$\therefore \delta t_{ik}^{l} = \left(\overset{II}{\Gamma^{r}_{is}} t^{l}_{rk} + \overset{II}{\Gamma^{r}_{ks}} t^{l}_{ri} - \overset{I}{\Gamma^{l}_{rs}} t^{t}_{ik} \right) \delta\xi^{s}. \tag{4.3.10}$$

Similarly, applying the law (4.3.10) to the parallel displacement of the Kronecker tensor, we get

$$\delta_{l}^{k} = a^{k} a_{i} \quad \therefore \; \delta\left(\delta_{l}^{k}\right) = \left(\overset{II}{\Gamma^{r}_{is}} - \overset{I}{\Gamma^{r}_{is}} \right) \delta\xi^{s}. \tag{4.3.11a}$$

For

$$\delta\left(\delta_{i}^{k}\right) = \delta\left(a^{k} a_{i}\right) \qquad \because \delta_{i}^{k} = a^{k} a_{i}$$

$$= \delta a^{k} a_{i} + a^{k} \delta a_{i}$$

$$= -\overset{I}{\Gamma^{k}_{rs}} a^{r} a_{i} \delta\xi^{s} + a^{k} \overset{II}{\Gamma^{r}_{is}} a_{r} \delta\xi^{s}$$

$$= \left(-\overset{I}{\Gamma^{k}_{rs}} \delta_{i}^{r} + \overset{II}{\Gamma^{r}_{is}} \delta_{r}^{k} \right) \delta\xi^{s} \tag{4.3.11b}$$

$$= \left(\overset{II}{\Gamma^{k}_{is}} - \overset{I}{\Gamma^{k}_{is}} \right) \delta\xi^{s}$$

$$\therefore \delta\left(\delta_{i}^{k}\right) = \left(\overset{II}{\Gamma^{k}_{is}} - \overset{I}{\Gamma^{k}_{is}} \right) \delta\xi^{s}.$$

Now, we apply Equation (4.3.11a) to the parallel displacement of the product $a^{i} \delta_{i}^{k}$ so that

$$\delta\left(a^i\delta_i^k\right)=\delta_i^k\delta a^i+a^i\delta\left(\delta_i^k\right)$$

$$=\delta a^k+a^i\delta\left(\delta_i^k\right)$$

$$\therefore\ \delta a^k=\delta a^k+a^i\delta\left(\delta_i^k\right)$$

$$\Rightarrow a^i\delta\left(\delta_i^k\right)=0$$

$$\Rightarrow \delta\left(\delta_i^k\right)=0 \qquad \left(\because a^i \text{ is arbitrary}\right).$$

Now (4.3.11b) reduces to $\left(\overset{II}{\Gamma}{}_{is}^{k}-\overset{I}{\Gamma}{}_{is}^{k}\right)\delta\xi^s=0$

$$\therefore \overset{II}{\Gamma}{}_{is}^{k}=\overset{I}{\Gamma}{}_{is}^{k} \qquad \left(\because \delta\xi^s \text{ is arbitrary}\right).$$

Therefore, we shall omit the **distinguishing marks I and II**. The $\overset{I}{\Gamma}{}_{is}^{k}$ are symmetric in their subscripts if it is possible to introduce a coordinate system in which they vanish at least locally. Henceforth, we shall consider only symmetric $\overset{I}{\Gamma}{}_{is}^{k}$. The $\overset{I}{\Gamma}{}_{is}^{k}$ still to a high degree is arbitrary. They are, however, uniquely determined, if we connect them with the **metric tensor** g_{ik} by the following condition. The result of the **parallel displacement** of a vector $\vec{a}$ shall not depend on whether we apply the law of parallel displacement to its contravariant or covariant representation. The two representations of a^i and a_k, the components $\left(a^i+\delta a^i\right)$ and $\left(a_k+\delta a_k\right)$ at the point $\left(\xi^s+\delta\xi^s\right)$, respectively, where δa^i and δa_k are given by Equations (4.3.4) and (4.3.5). These two vectors are again the representations of the same vector $\left(a_k+\delta a_k\right)$ at the point $\left(\xi^s+\delta\xi^s\right)$, expressed equivalently by the equation:

$$a_k+\delta a_k=\left(g_{ik}+\delta g_{ik}\right)\left(a^i+\delta a^i\right), \tag{4.3.12}$$

where δg_{ik} is

$$\delta g_{ik}=g_{ik,l}\delta\xi^l.$$

Equation (4.3.12) must be satisfied up to **linear terms** in the differentials and for arbitrary a^i and $\delta\xi^s$. If we multiply the R.H.S. of (4.3.12), we obtain

$$\delta a_k=g_{ik}\delta a^i+\delta g_{ik}a^i$$

$$=g_{ik}\delta a^i+g_{ik,l}\delta\xi^i a^i.$$

Substituting δa^i and δa_k from Equations (4.3.4) and (4.3.5), we get

$$\Gamma^i_{kl}\, a_i \delta\xi^l = -g_{ik}\,\Gamma^i_{sl}\, a^s \delta\xi^l + g_{ik,l}\delta\xi^l a^i \qquad (i \to s)$$

$$g_{is} a^s \Gamma^i_{kl}\, \delta\xi^l + g_{ik}\Gamma^i_{sl}\, a^s \delta\xi^l - g_{sk,l} a^s \delta\xi^l = 0 \qquad \because a_i = g_{is} a^s \tag{4.3.13}$$

$$\Rightarrow \quad a^s \delta\xi^l \left[g_{is}\Gamma^i_{kl} + g_{ik}\Gamma^i_{sl} - g_{sk,i} \right] = 0.$$

Since a^s and $\delta\xi^l$ are arbitrary, the contents of the bracket **must vanish**.

4.4 Use of Symmetry Condition for the Ultimate Result

We make use of the symmetry condition and write down the vanishing bracket three times with different index combinations:

$$\Gamma^r_{ik} g_{rs} + \Gamma^r_{sk} g_{ir} - g_{is,k} = 0 \tag{i}$$

$$\Gamma^r_{ki} g_{rs} + \Gamma^r_{si} g_{rk} - g_{ks,i} = 0 \tag{ii}$$

$$\Gamma^r_{is} g_{rk} + \Gamma^r_{sk} g_{ir} - g_{ik,s} = 0 \tag{iii}$$

Now, (i) + (ii) – (iii) gives

$$-g_{is,k} - g_{ks,i} + g_{ik,s} + \Gamma^r_{ik} g_{rs} + \Gamma^r_{ik} g_{rs} = 0$$

$$\Rightarrow g_{is,k} + g_{ks,i} - g_{ik,s} = 2\Gamma^r_{ik} g_{rs}$$

$$\Rightarrow \frac{1}{2}\left[g_{is,k} + g_{ks,i} - g_{ik,s} \right] = \Gamma^r_{ik} g_{rs}. \tag{4.4.1}$$

Multiplying (4.4.1) by g^{sl}, we get

$$\delta^l_r \Gamma^r_{ik} = \frac{1}{2} g^{ls}\left[g_{is,k} + g_{ks,i} - g_{ik,s} \right]$$

$$\Rightarrow \quad \Gamma^l_{ik} = \frac{1}{2} g^{ls}\left[g_{is,k} + g_{ks,i} - g_{ik,s} \right]. \tag{4.4.2}$$

This expression is usually referred to as the Christoffel three-index symbol of the second kind, and it is denoted by $\left\{ \begin{matrix} l \\ ik \end{matrix} \right\}$ or Γ^l_{ik}.

$$\left\{ \begin{matrix} l \\ ik \end{matrix} \right\} = \frac{1}{2} g^{ls}\left\{ g_{is,k} + g_{ks,i} - g_{ik,s} \right\} = \Gamma^l_{ik}.$$

The L.H.S. of Equation (4.4.1) is called the Christoffel symbol of the first kind. It is denoted by the sign $[ik, s]$:

$$[ik,s]=\frac{1}{2}g^{is}\left[g_{is,k}+g_{ks,i}-g_{ik,s}\right]=\Gamma_{ik,s}.$$

In case of Cartesian coordinates, both kinds of Christoffel three-index symbols vanish, since g_{ij} = constant. Mathematically, the Christoffel symbols may be developed in this way from the **concept of parallel displacement**.

4.5 Coordinate Transformations of Christoffel Symbols

4.5.1 Transformation of the First Kind $\Gamma_{ij,k}$

Let g_{ij} and g'_{ij} be the fundamental tensors of the coordinate systems x^i and x'^i, respectively.

∴ By tensor law of transformation,

$$g'_{ij}=g_{ab}\frac{\partial x^a}{\partial x'^i}\frac{\partial x^b}{\partial x'^j} \qquad \begin{matrix} i & j & k \\ a & b & c \end{matrix}.$$

Differentiating partially both sides with respect to x'^k,

$$\frac{\partial g'_{ij}}{\partial x'^k}=\left(\frac{\partial g_{ab}}{\partial x^c}\cdot\frac{\partial x^c}{\partial x'^k}\right)\frac{\partial x^a}{\partial x'^i}\frac{\partial x^b}{\partial x'^j}+g_{ab}\frac{\partial^2 x^a}{\partial x'^i\partial x'^k}\frac{\partial x^b}{\partial x'^j}$$
$$+\underset{(b\leftrightarrow a)}{g_{ab}\frac{\partial x^a}{\partial x'^j}\frac{\partial^2 x^b}{\partial x'^j\partial x'^k}}. \tag{4.5.1}$$

Also from

$$g'_{jk}=g_{bc}\frac{\partial x^b}{\partial x'^j}\frac{\partial x^c}{\partial x'^k}$$

$$\frac{\partial g'_{jk}}{\partial x'^i}=\left(\frac{\partial g_{bc}}{\partial x^a}\cdot\frac{\partial x^a}{\partial x'^i}\right)\frac{\partial x^b}{\partial x'^j}\frac{\partial x^c}{\partial x'^k}+\underset{(b\to a,c\to b)}{g_{bc}\frac{\partial^2 x^b}{\partial x'^j\partial x'^i}\frac{\partial x^c}{\partial x'^k}} \tag{4.5.2}$$
$$+\underset{(c\to a)''}{g_{bc}\frac{\partial x^b}{\partial x'^j}\frac{\partial^2 x^c}{\partial x'^k\partial x'^i}}.$$

(differentiating partially both sides with respect to x'^i)

Similarly, from

$$g'_{ki} = g_{ca}\frac{\partial x^c}{\partial x'^k}\frac{\partial x^a}{\partial x'^i}$$

$$\frac{\partial g'_{ki}}{\partial x'^j} = \left(\frac{\partial g_{ca}}{\partial x^b}\frac{\partial x^b}{\partial x'^j}\right)\frac{\partial x^c}{\partial x'^k}\frac{\partial x^a}{\partial x'^i} + g_{ca}\underset{(c\to a,a\to b)}{\frac{\partial^2 x^c}{\partial x'^k \partial x'^j}\frac{\partial x^a}{\partial x'^i}} + g_{ca}\underset{(c\to b)''}{\frac{\partial x^c}{\partial x'^k}\frac{\partial^2 x^a}{\partial x'^i \partial x'^j}}. \tag{4.5.3}$$

(differentiating partially both sides with respect to x'^j)

(The indices are changed so as to convert all second-order derivatives in terms of x^a like $\frac{\partial^2 x^a}{\partial x'^i \partial x'^j}$ and indices of g to g_{ab}.)

(4.5.3) + (4.5.2) – (4.5.1) gives

$$\left(\frac{\partial g'_{ki}}{\partial x'^j} + \frac{\partial g_{jk}}{\partial x^i} - \frac{\partial g_{ij}}{\partial x'^k}\right)$$

$$= \left(\frac{\partial g_{ca}}{\partial x^b} + \frac{\partial g_{bc}}{\partial x^a} - \frac{\partial g_{ab}}{\partial x^c}\right)\frac{\partial x^a}{\partial x'^i}\frac{\partial x^b}{\partial x'^j}\frac{\partial x^c}{\partial x'^k} + 2g_{ab}\frac{\partial^2 x^a}{\partial x'^i \partial x'^j}\frac{\partial x^b}{\partial x'^k}$$

$$\Gamma'_{ij,k} = \Gamma_{ab,c}\frac{\partial x^a}{\partial x'^i}\frac{\partial x^b}{\partial x'^j}\frac{\partial x^c}{\partial x'^k} + g_{ab}\frac{\partial^2 x^a}{\partial x'^i \partial x'^j}\frac{\partial x^b}{\partial x'^k}. \tag{4.5.4}$$

This is the transformation law of the Christoffel symbol of the first kind. Clearly, due to the presence of the second term in (4.5.4), it is not the transformation law of a tensor; hence, the Christoffel bracket of the first kind $\Gamma_{ij,k}$ is **not a tensor in general.**

But, if we consider a linear transformation of the type

$$x^a = A_i^a x'^i + B^a \quad (a = 1, 2, \ldots, n),$$

then $\frac{\partial^2 x^a}{\partial x'^i \partial x'^j} = 0$ for which $\Gamma'_{ij,k} = \Gamma_{ab,c}\frac{\partial x^a}{\partial x'^i}\frac{\partial x^b}{\partial x'^j}\frac{\partial x^c}{\partial x'^k}$, which is the transformation law of a third rank covariant tensor and subject to the linear transformation; Christoffel symbol (bracket) of the first kind is also a tensor.

4.5.2 Transformation of the Second Kind Γ^i_{jk}

From the tensor law of transformation of the second fundamental tensor, we can write

$$g'^{pk} = g^{\alpha\beta}\frac{\partial x'^p}{\partial x^\alpha}\frac{\partial x'^k}{\partial x^\beta}$$

Multiplying (4.5.4) by this relation, we can get

$$g'^{pk}\Gamma'_{ij,k} = g^{\alpha\beta}\Gamma_{ab,c}\frac{\partial x^a}{\partial x'^i}\frac{\partial x^b}{\partial x'^j}\frac{\partial x^c}{\partial x'^k}\frac{\partial x'^p}{\partial x^\alpha}\frac{\partial x'^k}{\partial x^\beta} + g_{ab}g^{\alpha\beta}\frac{\partial^2 x^a}{\partial x'^i \partial x'^j}\frac{\partial x^b}{\partial x'^k}\frac{\partial x'^p}{\partial x^\alpha}\frac{\partial x'^k}{\partial x^\beta}$$

$$= g^{\alpha\beta}\Gamma_{ab,c}\frac{\partial x^a}{\partial x'^i}\frac{\partial x^b}{\partial x'^j}\delta^c_\beta\frac{\partial x'^p}{\partial x^\alpha} + g_{ab}g^{\alpha\beta}\frac{\partial^2 x^a}{\partial x'^i \partial x'^j}\delta^b_\beta\frac{\partial x'^p}{\partial x^\alpha}$$

$$= g^{\alpha c}\Gamma_{ab,c}\frac{\partial x^a}{\partial x'^i}\frac{\partial x^b}{\partial x'^j}1\cdot\frac{\partial x'^p}{\partial x^\alpha} + g_{ab}g^{\alpha b}\frac{\partial^2 x^a}{\partial x'^i \partial x'^j}1\cdot\frac{\partial x'^p}{\partial x^\alpha}$$

$$\therefore \Gamma'^p_{ij} = \Gamma^\alpha_{ab}\frac{\partial x^a}{\partial x'^i}\frac{\partial x^b}{\partial x'^j}\frac{\partial x'^p}{\partial x^\alpha} + \delta^\alpha_a\frac{\partial^2 x^a}{\partial x'^i \partial x'^j}\frac{\partial x'^p}{\partial x^\alpha}$$

$$\Gamma'^P_{ij} = \Gamma^\alpha_{ab}\frac{\partial x^a}{\partial x'^i}\frac{\partial x^b}{\partial x'^j}\frac{\partial x'^p}{\partial x^\alpha} + \frac{\partial^2 x^\alpha}{\partial x'^i \partial x'^j}\frac{\partial x'^p}{\partial x^\alpha}, \tag{4.5.5}$$

which is the transformation law of the Christoffel symbol Γ^P_{ij} of the second kind, and it is also not the tensor due to the presence of the second term in general.

For the linear transformation of the type,

$$x^\alpha = A^\alpha_i x'^i + \beta^\alpha, \quad \frac{\partial^2 x^\alpha}{\partial x'^i \partial x'^j} = 0.$$

From (4.5.5), we get

$\Gamma'^P_{ij} = \Gamma^\alpha_{ab}\frac{\partial x^a}{\partial x'^i}\frac{\partial x^b}{\partial x'^j}\frac{\partial x'^p}{\partial x^\alpha}$, which is the transformation law of a third rank mixed tensor. Hence, in this sense, the Christoffel symbol of the second kind is also a tensor.

Note: (4.5.5) gives $\Gamma'^P_{ij}\frac{\partial x^\alpha}{\partial x'^p} = \Gamma^\alpha_{ab}\frac{\partial x^a}{\partial x'^i}\frac{\partial x^b}{\partial x'^j} + \frac{\partial^2 x^\alpha}{\partial x'^i \partial x'^j}$, (Corollary 1) which will be of **immense use** in subsequent development.

Example 1

If the metric in a V_n is such that
$g_{ij} = 0$ for $i \neq j$, show that for all unequal i, j, k

a. $\Gamma^i_{jk} = 0.$

b. $\Gamma^i_{ij} = \frac{1}{2g_{ii}}\frac{\partial g_{ii}}{\partial x^j}$.

c. $\Gamma^{i}_{jj} = -\dfrac{1}{2g_{ii}}\dfrac{\partial g_{jj}}{\partial x^{i}}$.

d. $\Gamma^{i}_{ii} = \dfrac{1}{2g_{ii}}\dfrac{\partial g_{ii}}{\partial x^{i}}$.

Proof: Given $g_{ij} = 0$ for $i \neq j$ in a space V_n; $g_{ii} \neq 0$, i.e., g_{ii} exists. Therefore, $g^{ii} = \dfrac{l}{g_{ii}}$ also exists.

By definition,

a. $\Gamma^{i}_{jk} = g^{i\alpha}\Gamma_{jk,\alpha} = g^{ii}\Gamma_{jk,i} = \dfrac{1}{2g^{ii}}\left[\dfrac{\partial g_{ji}}{\partial x^{k}} + \dfrac{\partial g_{ki}}{\partial x_{j}} - \dfrac{\partial g_{jk}}{\partial x^{i}}\right] = 0$

$\because g_{ij} = 0$ for $i \neq j$.

b. $\Gamma^{i}_{ij} = g^{i\alpha}\Gamma_{ij,\alpha} = g^{ii}\Gamma_{ij,i} = \dfrac{1}{2g_{ii}}\left[\dfrac{\partial g_{ii}}{\partial x^{j}} + \dfrac{\partial g_{ji}}{\partial x^{i}} - \dfrac{\partial g_{ij}}{\partial x^{i}}\right]$

$= \dfrac{1}{2g_{ii}}\dfrac{\partial g_{ii}}{\partial x^{j}}\,(i \neq j)$ (with two distinct indices).

c. $\Gamma^{i}_{jj} = g^{i\alpha}\Gamma_{jj,\alpha} = g^{ii}\Gamma_{jj,i}$

$= \dfrac{1}{2g_{ii}}\left[\dfrac{\partial g_{ji}}{\partial x^{j}} + \dfrac{\partial g_{ji}}{\partial x^{j}} - \dfrac{\partial g_{jj}}{\partial x^{i}}\right]$

$= -\dfrac{1}{2g_{ii}}\dfrac{\partial g_{jj}}{\partial x^{i}}$ (with two distinct indices).

d. $\Gamma^{i}_{ii} = g^{i\alpha}\Gamma_{ii,\alpha} = g^{ii}\Gamma_{ii,i} = \dfrac{1}{2g_{ii}}\dfrac{\partial g_{ii}}{\partial x^{i}}$ (with one distinct index).

Hence, proved.

N.B.: These four results (a)–(d) will be of great help in determining nonvanishing Christoffel symbols of the second kind in future for many investigations.

Example 2

Find the nonvanishing Christoffel symbols of the second kind for the metric:

$$ds^2 = a^2\left(d\theta^2 + \sin^2\theta d\phi^2\right).$$

In this case, it is a metric in v_2 with two variables $\theta = x^1$ (say) and $\phi = x^2$ (say) so that

$$g_{11} = a^2, g_{12} = g_{21} = 0,\ g_{22} = a^2 \sin^2\theta \text{ and}$$

$$g = \begin{vmatrix} a^2 & 0 \\ 0 & a^2 \sin^2\theta \end{vmatrix} = a^4 \sin^2\theta \neq 0$$

$$\therefore\ \frac{\partial g_{11}}{\partial x^i} = 0\ \forall i,\ \frac{\partial g_{22}}{\partial \theta} = \frac{\partial g_{22}}{\partial x^1} = 2a^2 \sin\theta\cos\theta,\ \frac{\partial g_{22}}{\partial \phi} = 0.$$

a. For distinct i, j, k $(i, j, k = 1, 2)$,

$$\Gamma^i_{jk} = 0 \qquad \because g_{ij} = 0 \text{ for } i \neq j.$$

b. For two distinct indices,

$$\Gamma^i_{ij} = \frac{1}{2g_{ii}} \frac{\partial g_{ii}}{\partial x^j}\ (i \neq j).$$

$$\text{For } i = 1,\ j = 2\ \ \Gamma^1_{12} = \frac{1}{2g_{11}} \frac{\partial g_{11}}{\partial x^2} = 0.$$

$$\text{For } i = 2,\ j = 1\ \ \Gamma^2_{21} = \frac{1}{2g_{22}} \frac{\partial g_{22}}{\partial x^1}.$$

$$\therefore \Gamma^2_{21} = \frac{1}{2a^2 \sin^2\theta} \times 2a^2 \sin\theta \cos\theta = \cot\theta.$$

c. For (another) two distinct indices,

$$\Gamma^i_{jj} = \frac{-1}{2g_{ii}} \frac{\partial g_{jj}}{\partial x^i}.$$

$$\therefore\ \Gamma^1_{22} = -\frac{1}{2g_{11}} \frac{\partial g_{22}}{\partial x^1} = -\frac{1}{2a^2} \times 2a^2 \sin\theta\cos\theta$$

$$= -\sin\theta\cos\theta$$

$$\text{and } \Gamma^2_{11} = \frac{-1}{2g_{22}} \frac{\partial g_{11}}{\partial x^2} = 0.$$

d. With only one distinct index,

$$\Gamma^1_{11} = \frac{1}{2g_{11}} \frac{\partial g_{11}}{\partial x^1} = 0 \text{ and } \Gamma^2_{22} = \frac{1}{2g_{22}} \frac{\partial g_{22}}{\partial x^2} = 0.$$

Hence, the nonvanishing Christoffel symbols of the second kind are

$$\Gamma^2_{21} = \cot\theta,\ \Gamma^1_{22} = -\sin\theta\cos\theta.$$

We have already come across with the entity called tensors born out of non-isotropic medium, a set of functions which obey the transformation laws (2.2.2) and (2.2.3). It has already been mentioned that there are many fields such as viscous fluids, elasticity, structures prone to deformations, general theory of relativity, and continuum mechanics, where tensors are used. To study the nature of changes from the mathematical point of view, some variation concept parallel to the derivatives of functions in ordinary calculus needs to be developed applicable for tensors. We intend to develop this concept in Section 4.6.

4.6 Covariant Derivative of Covariant Tensor of Rank One

Let us consider a covariant tensor A_i of rank one, so by tensor law of transformation from x^i to x'^i:

$$A'_i = A_{\acute{a}} \frac{\partial x^{\acute{a}}}{\partial x'^i}.$$

Differentiating partially with respect to x'^j, we can get

$$\begin{aligned}
\frac{\partial A'_i}{\partial x'^j} &= \left(\frac{\partial A_a}{\partial x^b}\frac{\partial x^b}{\partial x'^j}\right)\frac{\partial x^{\acute{a}}}{\partial x'^i} + A_a \frac{\partial^2 x^a}{\partial x'^i \partial x'^j} \\
&= \frac{\partial A_a}{\partial x^b}\frac{\partial x^a}{\partial x'^i}\frac{\partial x^b}{\partial x'^j} + A_a\left[\Gamma'^p_{ij}\frac{\partial x^a}{\partial x'^p} - \Gamma^a_{\lambda\mu}\frac{\partial x^\lambda}{\partial x'^i}\frac{\partial x^\mu}{\partial x'^j}\right], \text{ using Corollary 1} \\
&= \frac{\partial A_a}{\partial x^b}\frac{\partial x^a}{\partial x'^i}\frac{\partial x^b}{\partial x'^j} + \left(A_a\frac{\partial x^a}{\partial x'^p}\right)\Gamma'^p_{ij} - A_a\Gamma^a_{\lambda\mu}\underset{\lambda\leftrightarrow a,\,\mu\to b}{\frac{\partial x^\lambda}{\partial x'^i}\frac{\partial x^\mu}{\partial x'^j}} \\
&= \left(\frac{\partial A_a}{\partial x^b} - A_\lambda\Gamma^\lambda_{ab}\right)\frac{\partial x^a}{\partial x'^i}\frac{\partial x^b}{\partial x'^j} + A'_p\Gamma'^p_{ij}
\end{aligned} \tag{4.6.1}$$

(The dummy indices are changed looking to the indices of the first term.)

$$\begin{aligned}
\frac{\partial A'_i}{\partial x'^j} - A'_p\Gamma'^p_{ij} &= \left(\frac{\partial A_a}{\partial x^b} - A_\lambda\Gamma^\lambda_{ab}\right)\frac{\partial x^a}{\partial x'^i}\frac{\partial x^b}{\partial x'^j} \\
A'_{i,j} &= A_{a,b}\frac{\partial x^a}{\partial x'^i}\frac{\partial x^b}{\partial x'^j}
\end{aligned} \tag{4.6.2}$$

writing

$$A_{a,b} = \frac{\partial A_a}{\partial x^b} - A_\lambda\Gamma^\lambda_{ab} \tag{4.6.3}$$

It we define the x^b-covariant derivative of A_a with respect to the fundamental tensor g_{ij} by means of (4.6.3) and denote it as $A_{a,b}$, the relation (4.6.2) corresponding to it represents the transformation law of a covariant tensor of rank two. The covariant derivative defined by (4.6.3) contains the first-order ordinary derivative $\frac{\partial A_a}{\partial x^b}$ following which the symbol $A_{a,b}$ is used for **"Covariant Derivative."**

N.B.: The first covariant derivative of covariant vector or tensor of rank one is found to increase the rank of the tensor by one.

4.7 Covariant Derivative of Contravariant Tensor of Rank One

Consider the contravariant tensor A^i of rank one. Therefore, by the transformation law of tensors,

$$A^i = \frac{\partial x^i}{\partial x'^a} A'^a$$

Differentiating partially with respect to x^j

$$\frac{\partial A^i}{\partial x^j} = \frac{\partial A'^a}{\partial x'^b}\frac{\partial x^i}{\partial x'^a}\frac{\partial x'^b}{\partial x^j} + A'^a \frac{\partial x'^b}{\partial x^j}\frac{\partial^2 x^i}{\partial x'^a \partial x'^b}$$

$$= \frac{\partial A'^a}{\partial x'^b}\frac{\partial x^i}{\partial x'^a}\frac{\partial x'^b}{\partial x^j} + A'^a \frac{\partial x'^b}{\partial x^j}\left[\Gamma'^p_{ab}\frac{\partial x^i}{\partial x'^p} - \Gamma^i_{\lambda\mu}\frac{\partial x^\lambda}{\partial x'^a}\frac{\partial x^\mu}{\partial x'^b}\right], \text{ using Corollary 1}$$

$$= \frac{\partial A'^a}{\partial x'^b}\frac{\partial x^i}{\partial x'^a}\frac{\partial x'^b}{\partial x^j} + A'^a \Gamma'^p_{ab}\underset{(p\leftrightarrow a)}{\frac{\partial x^i}{\partial x'^p}\frac{\partial x'^b}{\partial x^j}} - A'^a \frac{\partial x'^b}{\partial x^j}\frac{\partial x^\lambda}{\partial x'^a}\frac{\partial x^\mu}{\partial x'^b}\Gamma^i_{\lambda\mu} \tag{4.7.1}$$

$$= \frac{\partial A'^a}{\partial x'^b}\frac{\partial x^i}{\partial x'^a}\frac{\partial x'^b}{\partial x^j} + A'^p \Gamma'^a_{pb}\frac{\partial x^i}{\partial x'^a}\frac{\partial x'^b}{\partial x^j} - A'^a \frac{\partial x^\lambda}{\partial x'^a}\left(\frac{\partial x'^b}{\partial x^j}\frac{\partial x^\mu}{\partial x'^b}\right)\Gamma^i_{\lambda\mu}$$

$$= \left(\frac{\partial A'^a}{\partial x'^b} + A'^p \Gamma'^a_{pb}\right)\frac{\partial x^i}{\partial x'^a}\frac{\partial x'^b}{\partial x^j} - A^\lambda \Gamma^i_{\lambda\mu}\delta^\mu_j.$$

$$\therefore \frac{\partial A^i}{\partial x^j} + A^\lambda \Gamma^i_{\lambda j} = \left(\frac{\partial A'^a}{\partial x'^b} + A'^p \Gamma'^a_{pb}\right)\frac{\partial x^i}{\partial x'^a}\frac{\partial x'^b}{\partial x^j} \tag{4.7.2}$$

Writing

$$A^i_{,j} = \frac{\partial A^i}{\partial x^j} + A^\lambda \Gamma^i_{\lambda j} \tag{4.7.3}$$

in the relation (4.7.2), we can get

$$A^i_{,j} = A'^a_{,b} \frac{\partial x^i}{\partial x'^a} \frac{\partial x'^b}{\partial x^j}, \tag{4.7.4}$$

which is the transformation law of a mixed tensor of rank two, and (4.7.3) is called the x^j-covariant derivative of the contravariant tensor A^i of rank one with respect to the fundamental tensor g_{ij}. As it satisfies the transformation law (4.7.4) of tensors, so it must be a tensor. For the presence of the ordinary derivative $\frac{\partial A^i}{\partial x^j}$ in the covariant derivative, defined in (4.7.3), it is symbolically denoted by $A^i_{,j}$.

Note 1: The consideration of covariant derivative of A^i is also found to increase its rank by one.

Note 2: Due to the presence of the second term in (4.7.1), the partial derivative of the tensor $\frac{\partial A^i}{\partial x^j}$ is not a tensor in general. When can it be a tensor? It is left for guessing.

4.8 Covariant Derivative of Covariant Tensor of Rank Two

Consider the covariant tensor A_{ij} of rank two.

Therefore, by transformation law of tensors,

$$A'_{ij} = A_{ab} \frac{\partial x^a}{\partial x'^i} \frac{\partial x^b}{\partial x'^j}.$$

Differentiating partially with respect to x'^k, we get

$$\frac{\partial A'_{ij}}{\partial x'^k} = \left(\frac{\partial A_{ab}}{\partial x^c} \cdot \frac{\partial x^c}{\partial x'^k} \right) \frac{\partial x^a}{\partial x'^i} \frac{\partial x^b}{\partial x'^j} + A_{ab} \frac{\partial^2 x^a}{\partial x'^i \partial x'^k} \frac{\partial x^b}{\partial x^j} + A_{ab} \frac{\partial x^a}{\partial x'^i} \frac{\partial^2 x^b}{\partial x'^j \partial x'^k}$$

$$= \frac{\partial A_{ab}}{\partial x^c} \frac{\partial x^a}{\partial x'^i} \frac{\partial x^b}{\partial x'^j} \frac{\partial x^c}{\partial x'^k} + A_{ab} \frac{\partial x^b}{\partial x'^j} \left[\Gamma'^p_{ik} \frac{\partial x^a}{\partial x'^p} - \Gamma^a_{\lambda\mu} \frac{\partial x^\lambda}{\partial x'^i} \frac{\partial x^\mu}{\partial x'^k} \right] \tag{4.8.1}$$

$$+ A_{ab} \frac{\partial x^a}{\partial x'^i} \left[\Gamma'^p_{jk} \frac{\partial x^b}{\partial x'^p} - \Gamma^b_{\lambda\mu} \frac{\partial x^\lambda}{\partial x'^j} \frac{\partial x^\mu}{\partial x'^k} \right].$$

(Replacing the second-order derivatives by the corresponding differences of Christoffel's brackets of the second kind from Corollary 1, $\Gamma_{ij}^{\prime p}\frac{\partial x^{\alpha}}{\partial x^{\prime p}}=\Gamma_{\lambda\mu}^{\alpha}\frac{\partial x^{\lambda}}{\partial x^{\prime i}}\frac{\partial x^{\mu}}{\partial x^{\prime j}}+\frac{\partial^2 x^{\alpha}}{\partial x^{\prime i}\partial x^{\prime j}}$.)

$$\frac{\partial A'_{ij}}{\partial x'^k}=\frac{\partial A_{ab}}{\partial x^c}\frac{\partial x^a}{\partial x'^i}\frac{\partial x^b}{\partial x'^j}\frac{\partial x^c}{\partial x'^k}+A_{ab}\frac{\partial x^b}{\partial x'^j}\frac{\partial x^a}{\partial x'^p}\Gamma_{ik}^{\prime p}-A_{ab}\Gamma_{\lambda\mu}^{a}\frac{\partial x^b}{\partial x'^j}\underbrace{\frac{\partial x^{\lambda}}{\partial x'^i}\frac{\partial x^{\mu}}{\partial x'^k}}_{(\lambda\leftrightarrow a,\mu\to c)}$$

$$+A_{ab}\frac{\partial x^a}{\partial x'^i}\frac{\partial x^b}{\partial x'^p}\Gamma_{jk}^{\prime p}-A_{ab}\Gamma_{\lambda\mu}^{b}\frac{\partial x^a}{\partial x'^i}\frac{\partial x^{\lambda}}{\partial x'^j}\frac{\partial x^{\mu}}{\partial x'^k}{}_{(\lambda\leftrightarrow b,\mu\to c)}$$

(The dummy indices are to be changed looking to the indices of the factors of the first term.)

$$=\frac{\partial A_{ab}}{\partial x^c}\frac{\partial x^a}{\partial x'^i}\frac{\partial x^b}{\partial x'^j}\frac{\partial x^c}{\partial x'^k}+A'_{pj}\Gamma_{ik}^{\prime p}-A_{\lambda b}\Gamma_{ac}^{\lambda}\frac{\partial x^a}{\partial x'^i}\frac{\partial x^b}{\partial x'^j}\frac{\partial x^c}{\partial x'^k}$$

$$+A'_{ip}\Gamma_{jk}^{\prime p}-A_{a\lambda}\Gamma_{bc}^{\lambda}\frac{\partial x^a}{\partial x'^i}\frac{\partial x^b}{\partial x'^j}\frac{\partial x^c}{\partial x'^k}$$

$$\therefore\frac{\partial A'_{ij}}{\partial x'^k}-A'_{pj}\Gamma_{ik}^{\prime p}-A'_{ip}\Gamma_{jk}^{\prime p}=\left(\frac{\partial A_{ab}}{\partial x^c}-A_{\lambda b}\Gamma_{ac}^{\lambda}-A_{a\lambda}\Gamma_{bc}^{\lambda}\right)\times\frac{\partial x^a}{\partial x'^i}\frac{\partial x^b}{\partial x'^j}\frac{\partial x^c}{\partial x'^k}$$

Writing

$$A_{ab,c}=\frac{\partial A_{ab}}{\partial x^c}-A_{\lambda b}\Gamma_{ac}^{\lambda}-A_{a\lambda}\Gamma_{bc}^{\lambda},\tag{4.8.2}$$

the above relation can be thrown to the form:

$$A'_{ij,k}=A_{ab,c}\frac{\partial x^a}{\partial x'^i}\frac{\partial x^b}{\partial x'^j}\frac{\partial x^c}{\partial x'^k}\tag{4.8.3}$$

The x^c-covariant derivative of A_{ab} with respect to the fundamental tensor g_{ij} is defined by (4.8.2), and (4.8.3) shows that it is a tensor of rank three. Hence, covariant derivative is found to increase the rank of the tensor A_{ab} by one.

Note: From (4.8.1), it is found that the partial derivative $\frac{\partial A'_{ij}}{\partial x'^k}$ of the tensor A'_{ij} is not a tensor due to the presence of the second two terms. Search when can it be a tensor?

4.9 Covariant Derivative of Contravariant Tensor of Rank Two

Consider the contravariant tensor A^{ij} of rank two.

By transformation law of tensors,

$$A^{ij} = A'^{ab} \frac{\partial x^i}{\partial x'^a} \frac{\partial x^j}{\partial x'^b} \qquad \left\{ \begin{matrix} i & j & k \\ a & b & c \end{matrix} \right. .$$

Differentiating partially with respect to x^k, we get

$$\frac{\partial A^{ij}}{\partial x^k} = \left(\frac{\partial A'^{ab}}{\partial x'^c} \frac{\partial x'^c}{\partial x^k} \right) \frac{\partial x^i}{\partial x'^a} \frac{\partial x^j}{\partial x'^b} + A'^{ab} \left(\frac{\partial^2 x^i}{\partial x'^a \partial x'^c} \frac{\partial x'^c}{\partial x^k} \right) \frac{\partial x^j}{\partial x'^b}$$

$$+ A'^{ab} \frac{\partial x^i}{\partial x'^a} \left(\frac{\partial^2 x^j}{\partial x'^b \partial x'^c} \frac{\partial x'^c}{\partial x^k} \right)$$

$$= \frac{\partial A'_{ab}}{\partial x'^c} \frac{\partial x^i}{\partial x'^a} \frac{\partial x^j}{\partial x'^b} \frac{\partial x'^c}{\partial x^k} + A'^{ab} \frac{\partial x'^c}{\partial x^k} \frac{\partial x^j}{\partial x'^b} \times \left[\Gamma'^{p}_{ac} \frac{\partial x^i}{\partial x'^p} - \Gamma^{i}_{\lambda\mu} \frac{\partial x^\lambda}{\partial x'^a} \frac{\partial x^\mu}{\partial x'^c} \right]$$

$$+ A'^{ab} \frac{\partial x^i}{\partial x'^a} \frac{\partial x'^c}{\partial x^k} \left[\Gamma'^{p}_{bc} \frac{\partial x^j}{\partial x'^p} - \Gamma^{j}_{\lambda\mu} \frac{\partial x^\lambda}{\partial x'^b} \frac{\partial x^\mu}{\partial x'^c} \right] \qquad (4.9.1)$$

$$= \frac{\partial A'^{ab} \partial x^i}{\partial x'^c \partial x'^a} \frac{\partial x^j}{\partial x'^b} \frac{\partial x'^c}{\partial x^k} + A'^{ab} \Gamma'^{p}_{ac} \underset{(p \leftrightarrow a)}{\frac{\partial x^i}{\partial x'^p} \frac{\partial x^j}{\partial x'^b}} \frac{\partial x'^c}{\partial x^k}$$

$$- A'^{ab} \Gamma^{i}_{\lambda\mu} \frac{\partial x^\lambda}{\partial x'^a} \frac{\partial x^j}{\partial x'^b} \left(\frac{\partial x'^c}{\partial x^k} \frac{\partial x^\mu}{\partial x'^c} \right)$$

$$+ \underset{(p \leftrightarrow b)}{A'^{ab}} \Gamma'^{p}_{bc} \frac{\partial x^i}{\partial x'^a} \frac{\partial x^j}{\partial x'^p} \frac{\partial x'^c}{\partial x^k} - A'^{ab} \Gamma^{j}_{\lambda\mu} \frac{\partial x^i}{\partial x'^a} \frac{\partial x^\lambda}{\partial x'^b} \left(\frac{\partial x'^c}{\partial x^k} \frac{\partial x^\mu}{\partial x'^c} \right).$$

(changing the dummy indices in light of the indices of the factors of the first term)

$$\frac{\partial A^{ij}}{\partial x^k} = \frac{\partial A'^{ab} \partial x^i}{\partial x'^c \partial x'^a} \frac{\partial x^j}{\partial x'^b} \frac{\partial x'^c}{\partial x^k} + A'^{pb} \Gamma'^{a}_{pc} \frac{\partial x^i}{\partial x'^a} \frac{\partial x^j}{\partial x'^b} \frac{\partial x'^c}{\partial x^k} - A'^{ab} \Gamma^{i}_{\lambda\mu} \frac{\partial x^\lambda}{\partial x'^a} \frac{\partial x^j}{\partial x'^b} \delta^\mu_k$$

$$+ A'^{ap} \Gamma'^{b}_{pc} \frac{\partial x^i}{\partial x'^a} \frac{\partial x^j}{\partial x'^b} \frac{\partial x'^c}{\partial x^k} - A'^{ab} \Gamma^{j}_{\lambda\mu} \frac{\partial x^i}{\partial x'^a} \frac{\partial x^\lambda}{\partial x'^b} \delta^\mu_k$$

$$= \left(\frac{\partial A'^{ab}}{\partial x'^c} + A'^{pb} \Gamma'^{a}_{pc} + A'^{ap} \Gamma'^{b}_{pc} \right) \frac{\partial x^i}{\partial x'^a} \frac{\partial x^j}{\partial x'^b} \frac{\partial x'^c}{\partial x^k} - A^{\lambda j} \Gamma^{i}_{\lambda k} \cdot 1 - A^{i\lambda} \Gamma^{j}_{\lambda k} \cdot 1.$$

$$\therefore \frac{\partial A^{ij}}{\partial x^k} + A^{\lambda j}\Gamma_{\lambda k}^{i} + A^{i\lambda}\Gamma_{\lambda k}^{j} = \left(\frac{\partial A'^{ab}}{\partial x'^c} + A'^{pb}\Gamma_{pc}'^{a} + A'^{ap}\Gamma_{pc}'^{b}\right) \times \frac{\partial x^i}{\partial x'^a}\frac{\partial x^j}{\partial x'^b}\frac{\partial x'^c}{\partial x^k}$$

$$A^{ij}_{,k} = A'^{ab}_{,c}\frac{\partial x^i}{\partial x'^a}\frac{\partial x^j}{\partial x'^b}\frac{\partial x'^c}{\partial x^k}. \tag{4.9.2}$$

Writing

$$A^{ij}_{,k} = \frac{\partial A^{ij}}{\partial x^k} + A^{\lambda j}\Gamma_{\lambda k}^{i} + A^{i\lambda}\Gamma_{\lambda k}^{j}. \tag{4.9.3}$$

$A^{ij}_{,k}$ of (4.9.3) is called the x^k-covariant derivative of the second-order contravariant tensor A^{ij} with respect to the fundamental tensor g_{ij}. The corresponding result (4.9.2) being the transformation law of a third-order mixed tensor must be a tensor. Hence, the covariant derivative thus defined by (4.9.3) is also a tensor, and due to consideration of covariant derivative, the rank of the tensor is found to increase by one.

4.10 Covariant Derivative of Mixed Tensor of Rank Two

Consider the second-order mixed tensor A^i_j.

$\therefore$ By transformation law of tensors,

$$A^i_j = A'^a_b\frac{\partial x^i}{\partial x'^a}\frac{\partial x'^b}{\partial x^j}.$$

Differentiating partially with respect to x^k, we can get

$$\frac{\partial A^i_j}{\partial x^k} = \left(\frac{\partial A'^a_b}{\partial x'^c}\frac{\partial x'^c}{\partial x^k}\right)\frac{\partial x^i}{\partial x'^a}\frac{\partial x'^b}{\partial x^j} + A'^a_b\left(\frac{\partial^2 x^i}{\partial x'^a x'^c}\frac{\partial x'^c}{\partial x^k}\right)\frac{\partial x'^b}{\partial x^j} + A'^a_b\frac{\partial x^i}{\partial x'^a}\frac{\partial^2 x'^b}{\partial x^j\partial x^k}$$

$$= \frac{\partial A'^a_b}{\partial x'^c}\frac{\partial x^i}{\partial x'^a}\frac{\partial x'^b}{\partial x^j}\frac{\partial x'^c}{\partial x^k} + A'^a_b\frac{\partial x'^b}{\partial x^j}\frac{\partial x'^c}{\partial x^k}\left(\Gamma_{ac}'^{p}\frac{\partial x^i}{\partial x'^p} - \Gamma_{\lambda\mu}^{i}\frac{\partial x^\lambda}{\partial x'^a}\frac{\partial x^\mu}{\partial x'^c}\right)$$

$$+ A'^a_b\frac{\partial x^i}{\partial x'^a}\left[\Gamma_{jk}^{p}\frac{\partial x'^b}{\partial x^p} - \Gamma_{\lambda\mu}'^{b}\frac{\partial x'^\lambda}{\partial x^j}\frac{\partial x'^\mu}{\partial x^k}\right],\ \text{using Corollary 1}$$

$$= \frac{\partial A_b'^a}{\partial x'^c}\frac{\partial x^i}{\partial x'^a}\frac{\partial x'^b}{\partial x^j}\frac{\partial x'^c}{\partial x^k} + A_b'^a \Gamma_{ac}'^p \underset{(p\leftrightarrow a)}{\frac{\partial x^i}{\partial x'^p}}\frac{\partial x'^b}{\partial x^j}\frac{\partial x'^c}{\partial x^k} + A_b'^a \frac{\partial x^i}{\partial x'^a}\frac{\partial x'^b}{\partial x^p}\Gamma_{jk}^p$$

$$- A_b'^a \Gamma_{\lambda\mu}^i \frac{\partial x'^b}{\partial x^j}\frac{\partial x^\lambda}{\partial x'^a}\left(\frac{\partial x'^c}{\partial x^k}\frac{\partial x^\mu}{\partial x'^c}\right) - A_b'^a \frac{\partial x^i}{\partial x'^a}\frac{\partial x'^\lambda}{\partial x^j}\frac{\partial x'^\mu}{\partial x^k}\Gamma_{\lambda\mu}'^b \; (\lambda \leftrightarrow b, \mu \rightarrow c)$$

$$\frac{\partial A_j^i}{\partial x^k} = \frac{\partial A_b'^a}{\partial x'^c}\frac{\partial x^i}{\partial x'^a}\frac{\partial x'^b}{\partial x^j}\frac{\partial x'^c}{\partial x^k} + A_b'^p \Gamma_{pc}'^a \frac{\partial x^i}{\partial x'^a}\frac{\partial x'^b}{\partial x^j}\frac{\partial x'^c}{\partial x^k}$$

$$+ A_p^i \Gamma_{jk}^p - A_b'^a \frac{\partial x^\lambda}{\partial x'^a}\frac{\partial x'^b}{\partial x^j}\Gamma_{\lambda\mu}^i \delta_k^\mu - A_\lambda'^a \Gamma_{bc}'^\lambda \frac{\partial x^i}{\partial x'^a}\frac{\partial x'^b}{\partial x^j}\frac{\partial x'^c}{\partial x^k}$$

$$\therefore \frac{\partial A_j^i}{\partial x^k} + A_j^\lambda \Gamma_{\lambda k}^i - A_p^i \Gamma_{jk}^p = \left(\frac{\partial A_b'^a}{\partial x'^c} + A_b'^p \Gamma_{pc}'^a - A_\lambda'^a \Gamma_{bc}'^\lambda\right) \times \frac{\partial x^i}{\partial x'^a}\frac{\partial x'^b}{\partial x^j}\frac{\partial x'^c}{\partial x^k}$$

$$\therefore A_{j,k}^i = A_{b,c}'^a \frac{\partial x^i}{\partial x'^a}\frac{\partial x'^b}{\partial x^j}\frac{\partial x'^c}{\partial x^k}. \tag{4.10.1}$$

Writing

$$A_{j,k}^i = \frac{\partial A_j^i}{\partial x^k} + A_j^\lambda \Gamma_{\lambda k}^i - A_p^i \Gamma_{jk}^p. \tag{4.10.2}$$

In this case, $A_{j,k}^i$ of (4.10.2) is called the x^k-covariant derivative of the mixed tensor A_j^i with respect to the fundamental tensor g_{ij}; consequently, (4.10.1) represents the corresponding transformation law of a mixed tensor of rank three. Hence, the covariant derivative thus defined by (4.10.2) of the mixed tensor is also found to increase the rank by one.

Note: In (4.10.2), when the covariant index j is associated with k (third term), the corresponding term is **negative**, and when the contravariant index i is associated with k, the corresponding term (second) is **positive**.

4.10.1 Generalization

Following the above results, the covariant derivative of higher order tensors can be written as

i. $A_{ijk,l} = \dfrac{\partial A_{ijk}}{\partial x^l} - A_{\alpha jk}\Gamma_{il}^\alpha - A_{i\alpha k}\Gamma_{jl}^\alpha - A_{ij\alpha}\Gamma_{kl}^\alpha.$

ii. $A_{,l}^{ijk} = \dfrac{\partial A^{ijk}}{\partial x^l} + A^{ajk}\Gamma_{al}^i + A^{iak}\Gamma_{al}^j + A^{ija}\Gamma_{al}^k.$

iii. $A^{ij}_{k,l} = \frac{\partial A^{ij}_k}{\partial x^l} - A^{ij}_{\alpha}\Gamma^{\alpha}_{kl} + A^{\alpha j}_{k}\Gamma^{i}_{\alpha l} + A^{i\alpha}_{k}\Gamma^{j}_{\alpha l}.$

iv. $A^{i}_{jk,l} = \frac{\partial A^{i}_{jk}}{\partial x^l} - A^{i}_{\alpha k}\Gamma^{\alpha}_{jl} - A^{i}_{j\alpha}\Gamma^{\alpha}_{kl} + A^{\alpha}_{jk}\Gamma^{i}_{\alpha l}.$

Similarly,

$$A^{ij...k}_{lm...n,r} = \frac{\partial}{\partial x^r}\left(A^{ij...k}_{lm...n}\right) + A^{aj...k}_{lm...n}\Gamma^{i}_{ar} + A^{ia...k}_{lm...n}\Gamma^{j}_{ar} + \cdots$$

$$+ A^{ij...a}_{lm...n}\Gamma^{k}_{ar} - A^{ij...k}_{bm...n}\Gamma^{b}_{lr} - A^{ij...k}_{lb...n}\Gamma^{b}_{mr} - \cdots - A^{ij...k}_{lm...b}\Gamma^{b}_{nr}.$$

The generating factor of any space is the metric $ds^2 = g_{ij}(x^i)dx^i dx^j$, which contains the metric functions $g_{ij}(x^i)$. But it is proved to be a tensor called fundamental tensor, and the evolution of the branch tensor (the evolution of the branch tensor analysis) is dependent on this fundamental concept. This demands the consideration of covariant derivatives of all the fundamental tensors g_{ij}, g^{ij} and also g^i_j.

4.11 Covariant Derivatives of g_{ij}, g^{ij} and also g^i_j

i. Using the result (4.8.2), the x^k-covariant derivative of g_{ij}, we can write

$$g_{ij,k} = \frac{\partial g_{ij}}{\partial x^k} - g_{\alpha j}\Gamma^{\alpha}_{ik} - g_{i\alpha}\Gamma^{\alpha}_{jk}$$

$$= \frac{\partial g_{ij}}{\partial x^k} - \left(\Gamma_{ik,j} + \Gamma_{jk,i}\right) \quad \because g_{\alpha j}\Gamma^{\alpha}_{ik} = \Gamma_{ik,j}$$

$$= \frac{\partial g_{ij}}{\partial x^k} - \frac{\partial g_{ij}}{\partial x^k},$$

using $\Gamma_{jk,i} + \Gamma_{ik,j} = \frac{\partial g_{ji}}{\partial x^k}$

$g_{ij,k} = 0.$

ii. It is already proved that $g_{ij}g^{jk} = \delta^k_i = 1$ or 0.
Differentiating partially with respect to x^l,

$\frac{\partial g_{ij}}{\partial x^l}g^{jk} + g_{ij}\frac{\partial g^{jk}}{\partial x^l} = 0$ for both the cases.

$$\therefore \left(\Gamma_{jl,i} + \Gamma_{li,j}\right)g^{jk} + g_{ij}\frac{\partial g^{jk}}{\partial x^l} = 0$$

$$\therefore g_{ij}\frac{\partial g^{jk}}{\partial x^l}+g^{jk}\Gamma_{jl,i}+\Gamma_{li}^{k}=0.$$

To make $\frac{\partial g^{jk}}{\partial x^l}$ free from g_{ij}, we need to consider the inner product of it with g^{mi}:

$$\therefore g^{mi}g_{ij}\frac{\partial g^{jk}}{\partial x^l}+g^{mi}g^{jk}\Gamma_{jl,i}+g^{mi}\Gamma_{li}^{k}=0$$

$$\therefore \delta_j^m\frac{\partial g^{jk}}{\partial x^l}+g^{jk}\left(g^{mi}\Gamma_{jl,i}\right)+g^{mi}\Gamma_{li}^{k}=0$$

or $1\cdot\frac{\partial g^{mk}}{\partial x^l}+g^{jk}\Gamma_{jl}^{m}+g^{mi}\Gamma_{il}^{k}=0$

$$\therefore g^{mk}_{,l}=0$$

This follows $g^{ij}_{,k}=0.$

iii. Following the result (4.10.2), the x^k-covariant derivative of g_j^i can be written as

$$g^i_{j,k}=\frac{\partial g^i_j}{\partial x^k}-g^i_\alpha\Gamma_{jk}^{\alpha}+g^\alpha_j\Gamma_{\alpha k}^{i}=0-1\cdot\Gamma_{jk}^{i}+1\cdot\Gamma_{jk}^{i}=0$$

$g^i_\alpha=1$ when $i=\alpha$
$\quad=0$ when $i\neq\alpha$.

Thus, it is found that $g_{ij,k}=0$, $g^{ij}_{,k}=0$, and $g^i_{j,k}=0$.

Since all the covariant derivatives of the fundamental tensors are zeros, they are called **covariant constants**.

4.12 Covariant Differentiations of Sum (or Difference) and Product of Tensors

a. Two tensors of the same rank and similar characters are conformable for addition or subtraction. So, consider the two tensors A_{ij} and B_{ij} each of rank two so that
   $A_{ij}+B_{ij}=C_{ij}$, a tensor of the same rank and character.

By the definition of x^k-covariant derivative of C_{ij} with respect to the fundamental tensor g_{ij},

$$\begin{aligned} C_{ij,k} &= \frac{\partial C_{ij}}{\partial x^k} - C_{\alpha j}\Gamma_{ik}^{\alpha} - C_{i\alpha}\Gamma_{jk}^{\alpha} \\ &= \frac{\partial}{\partial x^k}(A_{ij}+B_{ij}) - (A_{\alpha j}+B_{\alpha j})\Gamma_{ik}^{\alpha} - (A_{i\alpha}+B_{i\alpha})\Gamma_{jk}^{\alpha} \\ &= \left(\frac{\partial A_{ij}}{\partial x^k} - A_{\alpha j}\Gamma_{ik}^{\alpha} - A_{i\alpha}\Gamma_{jk}^{\alpha}\right) + \left(\frac{\partial B_{ij}}{\partial x^k} - B_{\alpha j}\Gamma_{ik}^{\alpha} - B_{i\alpha}\Gamma_{jk}^{\alpha}\right) \end{aligned} \tag{4.12.1}$$

$$(A_{ij}+B_{ij})_{,k} = A_{ij,k} + B_{ij,k}.$$

Hence, the x^k-covariant derivative of the sum of two tensors is equal to the sum of their covariant derivatives. This result will hold good for the sum of two or more tensors (when conformable) of any character covariant, contravariant, or mixed.

Similarly, it can be proved for the difference of two (or more) tensors:

$$(A_{ij}-B_{ij})_{,k} = A_{ij,k} - B_{ij,k}. \tag{4.12.2}$$

b. Consider any two tensors A^{ij}, B_k so that $A^{ij}B_k = C_k^{ij}$, an outer product.

Now the x^m-covariant derivative of C_k^{ij}, can be written by virtue of (4.10.2) as

$$\begin{aligned} C_{k,m}^{ij} &= \frac{\partial C_k^{ij}}{\partial x^m} - C_{\alpha}^{ij}\Gamma_{km}^{\alpha} + C_k^{\alpha j}\Gamma_{\alpha m}^{i} + C_k^{i\alpha}\Gamma_{\alpha m}^{j} \\ &= \frac{\partial}{\partial x^m}(A^{ij}B_k) - A^{ij}B_{\alpha}\Gamma_{km}^{\alpha} + A^{\alpha j}B_k\Gamma_{\alpha m}^{i} + A^{i\alpha}B_k\Gamma_{\alpha m}^{j} \quad \because A^{ij}B_k = C_k^{ij} \\ &= \frac{\partial A^{ij}}{\partial x^m}B_k + A^{ij}\frac{\partial B_k}{\partial x^m} - A^{ij}B_{\alpha}\Gamma_{km}^{\alpha} + A^{\alpha j}B_k\Gamma_{\alpha m}^{i} + A^{i\alpha}B_k\Gamma_{\alpha m}^{j} \\ &= \left(\frac{\partial A^{ij}}{\partial x^m} + A^{\alpha j}\Gamma_{\alpha m}^{i} + A^{i\alpha}\Gamma_{\alpha m}^{j}\right)B_k + A^{ij}\left(\frac{\partial B_k}{\partial x^m} - B_{\alpha}\Gamma_{km}^{\alpha}\right) \\ &= A_{,m}^{ij}B_k + A^{ij}B_{k,m} \end{aligned} \tag{4.12.3}$$

$$\therefore (A^{ij}B_k)_{,m} = A_{,m}^{ij}B_k + A^{ij}B_{k,m}$$

Thus, the covariant derivative of the product (outer) of two tensors obeys the product rule of ordinary derivatives.

Example 3

If $A^{ij}_{,k}$ is the x^k-covariant differentiation of the second-order contravariant tensor A^{ij} with respect to the fundamental tensor g_{ij}, prove that

$$A^{ij}_{,j} = \frac{1}{\sqrt{g}} \frac{\partial}{\partial x^j}\left(A^{ij}\sqrt{g}\right) + A^{jk}\Gamma^i_{jk}.$$

Also show that (i) the last term vanishes if A^{jk} is skew symmetric and

(ii) $A^j_{i,j} = \frac{1}{\sqrt{g}} \frac{\partial}{\partial x^j}\left(A^j_i\sqrt{g}\right) - A^j_k\Gamma^k_{ij}$.

Proof: By the definition of x^j-covariant derivative, we have

$$\therefore A^{ij}_{,j} = \frac{\partial A^{ij}}{\partial x^j} + A^{\alpha j}\Gamma^i_{\alpha j} + A^{i\alpha}\Gamma^j_{\alpha j}$$

(considering a contraction setting $j = k$ in $A^{ij}_{,k}$)

$$= \frac{\partial A^{ij}}{\partial x^j} + A^{\alpha j}\Gamma^i_{\alpha j} + A^{i\alpha}\frac{\partial}{\partial x^\alpha}\left(\log\sqrt{g}\right) \qquad (\alpha \to j), g > 0$$

$$= \frac{\partial A^{ij}}{\partial x^j} + A^{ij}\frac{1}{\sqrt{g}}\frac{\partial}{\partial x^j}\left(\sqrt{g}\right) + A^{kj}\Gamma^i_{kj} \; (\alpha \to k \text{ in the last term})$$

$$A^{ij}_{,j} = \frac{1}{\sqrt{g}}\frac{\partial}{\partial x^j}\left(A^{ij}\sqrt{g}\right) + A^{jk}\Gamma^i_{jk} \quad (j \leftrightarrow k).$$

Hence, proved.

i. $A^{jk}\Gamma^i_{jk} = -A^{kj}\Gamma^i_{jk} \quad \because A^{jk} = -A^{kj}$

$$= -A^{jk}\Gamma^i_{kj} \qquad (j \leftrightarrow k)$$

$$= -A^{jk}\Gamma^i_{jk} \quad \left(\because \Gamma^i_{jk} = \Gamma^i_{kj}\right)$$

$\therefore 2A^{jk}\Gamma^i_{jk} = 0 \quad \therefore A^{jk}\Gamma^i_{jk} = 0$

ii. Also, from the definition of x^k-covariant derivative,

$$A^j_{i,k} = \frac{\partial}{\partial x^k}\left(A^j_i\right) - A^j_\alpha\Gamma^\alpha_{ik} + A^\alpha_i\Gamma^j_{\alpha k}.$$

Considering a contraction with respect to j and k, i.e., setting $j = k$, we get

$$A^j_{i,j} = \frac{\partial}{\partial x^j}\left(A^j_i\right) - A^j_\alpha\Gamma^\alpha_{ij} + A^\alpha_i\Gamma^j_{\alpha j}$$

$$= \frac{\partial}{\partial x^j}\left(A^j_i\right) - \underset{(\alpha\to k)}{A^j_\alpha\Gamma^\alpha_{ij}} + A^\alpha_i\frac{\partial}{\partial x^\alpha}\underset{(\alpha\to j)}{\left((\log\sqrt{g}\right)}, \; g > 0$$

$$\therefore A^j_{i,j} = \frac{\partial}{\partial x^j}\left(A^j_i\right) - A^j_k \Gamma^k_{ij} + A^j_i \frac{1}{\sqrt{g}} \frac{\partial}{\partial x^j}\sqrt{g}$$

$$= \frac{1}{\sqrt{g}} \frac{\partial}{\partial x^j}\left(A^j_i \sqrt{g}\right) - A^j_k \Gamma^k_{ij}.$$

Hence, proved.

Example 4

If, at a specified point, the derivatives of the g_{ij}'s with respect to the coordinates are all zero, the components of covariant derivatives of a vector at the point are the same as ordinary derivatives.

Given that $\frac{\partial g_{ij}}{\partial x^k} = 0$ at a point P (say) $\forall\ i, j, k$.

If A_i is any covariant (may be contravariant also) vector, then its x^j-covariant derivative is given by

$$A_{i,j} = \frac{\partial A_i}{\partial x^j} - A_\alpha \Gamma^\alpha_{ij} = \frac{\partial A_i}{\partial x^j} - A_\alpha g^{\alpha\beta} \Gamma_{ij,\beta}$$

$$= \frac{\partial A_i}{\partial x^j} - A_\alpha g^{\alpha\beta} \frac{1}{2}\left[\frac{\partial g_{i\beta}}{\partial x^j} + \frac{\partial g_{j\beta}}{\partial x^i} - \frac{\partial g_{ij}}{\partial x^\beta}\right]$$

$$A_{i,j} = \frac{\partial A_i}{\partial x^j} \quad \because \frac{\partial g_{ij}}{\partial x^\beta} = 0 \quad \forall\ i, j, k \text{ at the point } P.$$

4.13 Gradient of an Invariant Function

The partial derivatives of an invariant function ϕ is defined as the components of a vector called grad ϕ or $\nabla\phi$.

Theorem of Invariant Function

If ϕ is invariant function, show that its covariant derivative is equal to its ordinary derivatives.

Proof: Let A_i be an arbitrary vector so that (ϕA_i) is also an arbitrary vector since ϕ is invariant.

$\therefore$ Its x^j-covariant derivative with respect to g_{ij} can be written as

$$(\phi A_i)_{,j} = \frac{\partial}{\partial x^j}(\phi A_i) - (\phi A)_\alpha \Gamma^\alpha_{ij}$$

$$= \phi \frac{\partial A_i}{\partial x^j} + A_i \frac{\partial \phi}{\partial x^j} - \phi A_\alpha \Gamma^\alpha_{ij}$$

$$\phi_{,j}A_i + \phi A_{i,j} = \phi\left(\frac{\partial A_i}{\partial x^j} - A_\alpha \Gamma_{ij}^{\alpha}\right) + A_i\frac{\partial \phi}{\partial x^j}$$

$$= \phi A_{i,j} + A_i\frac{\partial \phi}{\partial x^j}.$$

$$\therefore \phi_{,j}A_i - A_i\frac{\partial \phi}{\partial x^j} = 0$$

$$\therefore A_i\left(\phi_{,j} - \frac{\partial \phi}{\partial x^j}\right) = 0$$

$\therefore \phi_{,j} = \dfrac{\partial \phi}{\partial x^j}$ $\quad \because A_i$ is arbitrary.

$\therefore$ The x^j-covariant derivative of ϕ, namely, ϕ_j, is equal to its ordinary partial derivatives $\dfrac{\partial \phi}{\partial x^j}$.

Hence, these partial derivatives $\dfrac{\partial \phi}{\partial x^j}$ of the invariant function ϕ are the components of grade ϕ vector. Otherwise, the covariant derivative ϕ_j of the invariant function is a vector which is nothing but the $(\nabla\phi)$ vector.

4.14 Curl of a Vector

The x^j-covariant derivative of the covariant vector A_i with respect to the fundamental tensor g_{ij} is given by

$A_{i,j} = \dfrac{\partial A_i}{\partial x^j} - A_\alpha \Gamma_{ij}^{\alpha}$, which is a second-order covariant tensor.

The components $A_{i,j}$ is interpreted as obtained from A_i due to the operation "covariant differentiation" with respect to the fundamental tensor g_{ij}.

The curl of the vector $\bar{A}(A_i)$ is defined as

$$\text{Curl } \bar{A} = A_{i,j} - A_{j,i}$$

$$= \left(\frac{\partial A_i}{\partial x^j} - A_\alpha \Gamma_{ij}^{\alpha}\right) - \left(\frac{\partial A_j}{\partial x^i} - A_\alpha \Gamma_{ji}^{\alpha}\right)$$

$$= \frac{\partial A_i}{\partial x^j} - \frac{\partial A_j}{\partial x^i} \quad \because \Gamma_{ij}^{\alpha} = \Gamma_{ji}^{\alpha}$$

Hence, $\text{Curl } \bar{A}\left(= \text{Curl } A_i\right) = A_{i,j} - A_{j,i} = \dfrac{\partial A_i}{\partial x^j} - \dfrac{\partial A_j}{\partial x^i}.$

Theorem: The Necessary and Sufficient Condition That the First Covariant Derivative of a Covariant Vector Will Be Symmetric If the Vector Is Gradient of Some Invariant Function

Proof: By the definition of curl of a vector $\bar{A}$, we have

$$\text{Curl } \bar{A} = A_{i,j} - A_{j,i} = 0 \text{ if } A_{i,j} = A_{j,i} \text{ (symmetric).}$$

But curl $\nabla\phi = 0$, where ϕ is some scalar invariant.

Otherwise, if $\bar{A} = \nabla\phi$, $\because$ Curl $\bar{A} = 0$,

$\therefore$ Curl $\nabla\phi = 0$.

$$\therefore A_{i,j} - A_{j,i} = 0.$$

$$\therefore A_{i,j} = A_{j,i}, \text{ i.e., symmetric.}$$

Hence, proved.

4.15 Divergence of a Vector

The divergence of a contravariant vector $\bar{u}^{\alpha}(u^i)$ is defined as the result of contraction with respect to its covariant derivative.

In, $u^i_{,j} = \dfrac{\partial u^i}{\partial x^j} + u^{\alpha}\Gamma^i_{\alpha j}$

Let us allow the contraction setting $i = j$, so that

$$u^i_{,i} = \frac{\partial u^i}{\partial x^i} + u^{\alpha}\Gamma^i_{\alpha i} = \frac{\partial u^i}{\partial x^i} + u^{\alpha}\frac{\partial}{\partial x^{\alpha}}\left(\log\sqrt{g}\right), (g > 0)$$

$$(\alpha \to i)$$

$$= \frac{\partial u^i}{\partial x^i} + u^i\frac{\partial}{\partial x^i}\left(\log\sqrt{g}\right)$$

$$= \frac{\partial u^i}{\partial x^i} + u^i\frac{1}{\sqrt{g}}\frac{\partial}{\partial x^i}\left(\sqrt{g}\right)$$

$$= \frac{1}{\sqrt{g}}\frac{\partial}{\partial x^i}\left(u^i\sqrt{g}\right)$$

$$\therefore \text{div}\,\bar{u}\left(\text{or div}\,u^i\right) = u^i_{,i} = \frac{1}{\sqrt{g}}\frac{\partial}{\partial x^i}\left(\sqrt{g}u^i\right),$$

which is a scalar invariant.

4.16 Laplacian of a Scalar Invariant

From the expression of divergence of the contravariant vector $\bar{u}\left(u^i\right)$ (Section 4.15), we have

$$\therefore \operatorname{div}\bar{u}\left(=u^i_{,i}\right)=\frac{1}{\sqrt{g}}\frac{\partial}{\partial x^i}\left(\sqrt{g}u^i\right)$$

$$\therefore \operatorname{div}\bar{u}=\left(u^i_{,i}\right)=\frac{1}{\sqrt{g}}\frac{\partial}{\partial x^i}\left(\sqrt{g}\ g^{ij}u_j\right). \qquad \text{(i)}$$

But we have already proved that $\phi_{,j}=\frac{\partial\phi}{\partial x^j}$ (Section 2.13), which is the covariant derivative of the scalar invariant ϕ and is the component of $\nabla\phi$; (i) can be written as

$$\operatorname{div}(\nabla\phi)=\frac{1}{\sqrt{g}}\frac{\partial}{\partial x^i}\left(\sqrt{g}\ g^{ij}\phi_{,j}\right)$$

or

$$\nabla^2\phi=\frac{1}{\sqrt{g}}\frac{\partial}{\partial x^i}\left(\sqrt{g}\ g^{ij}\phi_{,j}\right). \qquad \text{(ii)}$$

Hence, the divergence of the vector $\nabla\phi$, symbolically $\nabla^2\phi$, is defined as the Laplacian of ϕ, and (ii) is its explicit expression.

In deriving div u^i from (i), we have replaced u^i as

$$u^i=g^{ij}u_j=g^{ij}\phi_{,j} \quad \because \operatorname{grad}\phi=\phi_{,j}=\frac{\partial\phi}{\partial x^j}$$

to recover the Laplacian of ϕ.

$u^i=g^{ij}\phi_{,j}$, since g^{ij} is symmetric.

Therefore, div $(\nabla\phi)$ can also be interpreted as the outcome of contraction of the covariant derivative of $g^{ij}\phi_{,j}$:

$$\therefore \nabla^2\phi=g^{ij}\phi_{,ij}$$

$$\because \phi_{,i}=\frac{\partial\phi}{\partial x^i} \text{ and } \operatorname{div}u^i=u^i_{,i}=(g^{ij}\phi_{,j})_{,i}$$

$$\nabla^2\phi=g^{ij}\left[\frac{\partial^2\phi}{\partial x^i\partial x^j}-\frac{\partial\phi}{\partial x^k}\Gamma^k_{ij}\right]$$

$$\phi_{,ij} = \frac{\partial}{\partial x^j}\left(\frac{\partial \phi}{\partial x^i}\right) - \frac{\partial \phi}{\partial x^k}\Gamma_{ij}^{k} \quad \because g^{ij}_{,i} = 0$$

and

$$\operatorname{div} u^i = u^i_{,i} = (g^{ij}u_j)_{,i}$$

$$= (g^{ij}\phi_{,j})_{,i} = g^{ij}\phi_{,ij},$$

which is another form of the Laplacian of ϕ.

Example 5

If A_{ij} is the curl of a covariant vector, prove that

$$A_{ij,k} + A_{jk,i} + A_{ki,j} = 0; \text{otherwise,}$$

$$\frac{\partial A_{ij}}{\partial x^k} + \frac{\partial A_{jk}}{\partial x^i} + \frac{\partial A_{ki}}{\partial x^j} = 0.$$

Let B_i be the covariant vector so that

$$A_{ij} = B_{i,j} - B_{j,i} = \frac{\partial B_i}{\partial x^j} - \frac{\partial B_j}{\partial x^i}. \tag{i}$$

$$A_{ji} = B_{j,i} - B_{i,j} \text{ interchanging } i \text{ and } j$$

$$= -\left(B_{i,j} - B_{j,i}\right)$$

$$= -A_{ij}.$$

Hence, A_{ij} is an antisymmetric tensor.

$$\therefore A_{ij} + A_{ji} = 0. \tag{ii}$$

Now,

$$A_{ij,k} + A_{jk,i} + A_{ki,j} = \frac{\partial A_{ij}}{\partial x^k} - A_{\alpha j}\Gamma_{ik}^{\alpha} - A_{i\alpha}\Gamma_{jk}^{\alpha}$$

$$+ \frac{\partial A_{jk}}{\partial x^i} - A_{\alpha k}\Gamma_{ji}^{\alpha} - A_{j\alpha}\Gamma_{ki}^{\alpha} + \frac{\partial A_{ki}}{\partial x^j} - A_{\alpha i}\Gamma_{kj}^{\alpha} - A_{k\alpha}\Gamma_{ij}^{\alpha}$$

$$= \frac{\partial A_{ij}}{\partial x^k} + \frac{\partial A_{jk}}{\partial x^i} + \frac{\partial A_{ki}}{\partial x^j} - \left(A_{\alpha j} + A_{j\alpha}\right)\Gamma_{ki}^{\alpha}$$

$$- \left(A_{i\alpha} + A_{\alpha i}\right)\Gamma_{jk}^{\alpha} - \left(A_{\alpha k} + A_{k\alpha}\right)\Gamma_{ij}^{\alpha}$$

$$= \frac{\partial A_{ij}}{\partial x^k} + \frac{\partial A_{jk}}{\partial x^i} + \frac{\partial A_{ki}}{\partial x^j}$$

using (ii) and $\Gamma_{ij}^{\alpha} = \Gamma_{ji}^{\alpha}$.

Also,

$$\frac{\partial A_{ij}}{\partial x^k} + \frac{\partial A_{jk}}{\partial x^i} + \frac{\partial A_{ki}}{\partial x^j} = \frac{\partial}{\partial x^k}\left(\frac{\partial B_i}{\partial x^j} - \frac{\partial B_j}{\partial x^i}\right) + \frac{\partial}{\partial x^i}\left(\frac{\partial B_j}{\partial x^k} - \frac{\partial B_k}{\partial x^j}\right)$$

$$+ \frac{\partial}{\partial x^j}\left(\frac{\partial B_k}{\partial x^i} - \frac{\partial B_i}{\partial x^k}\right), \quad \text{using (i)}$$

$$= \frac{\partial^2 B_i}{\partial x^k \partial x^j} - \frac{\partial^2 B_j}{\partial x^k \partial x^i} + \frac{\partial^2 B_j}{\partial x^i \partial x^k} - \frac{\partial^2 B_k}{\partial x^i \partial x^j}$$

$$+ \frac{\partial^2 B_k}{\partial x^j \partial x^i} - \frac{\partial^2 B_i}{\partial x^j \partial x^k} = 0.$$

Hence, proved.

4.17 Intrinsic Derivative or Derived Vector of $\vec{v}$

If $\hat{a}$ is any vector representing a direction and $\vec{v}$ is any vector with covariant components v_i (or contravariant components v^i), the intrinsic derivative or derived vector of $\vec{v}$ in the direction of $\hat{a}$ is defined by means of the covariant components $v_{i,k}a^k$ or by contravariant components $v^i_{,k}a^k$:

$$\left[\text{For } v_{i,k}a^k = \left(\delta^i_j\, v_j\right)_{,k} a^k = \left(g^{i\alpha} g_{j\alpha} v_j\right)_{,k} a^k \left(g^{i\alpha} v_\alpha\right)_{,K} a^k = v^i_{,k}a^k\right].$$

It is usually denoted by the symbol $\bar{a}.\nabla\bar{v}$, which is nothing but the generalization of the derivative of a vector in Euclidean space E_3 in the direction of $\bar{a}$.

Theorem: Show That a Vector of Constant Magnitude Is Orthogonal to Its Intrinsic Derivative

Proof: Let $\vec{u}(u_i)$ be a covariant vector of constant magnitude so that

$$u^2 = g^{ij}u_i u_j.$$

Considering the x^k-covariant differentiation with respect to the fundamental tensor g_{ij}, we can get

$$0 = g^{ij}\left(u_{i,k}u_j + u_i u_{j,k}\right) \quad \because g^{ij}_{,k} = 0$$

$(g^{ij}u_j)u_{i,k} + (g^{ij}u_j)u_{i,k} = 0$ ($i \leftrightarrow j$ in the second term)

$$u^i u_{i,k} + u^i u_{i,k} = 0 \therefore 2u^i u_{i,k} = 0$$

$u^i\ u_{i,k}\ a^k = 0$ (considering the inner product with a^k)

$\therefore u^i\left(u_{i,k}a^k\right) = 0$, which is of the form $u^i u_i = 0$.

This shows that the intrinsic derivative of $u_{i,k}$ in the direction of $\hat{a}\left(a^k\right)$ is orthogonal to $\bar{u}$ itself.

Hence, proved.

4.18 Definition: Parallel Displacement of Vectors

4.18.1 When Magnitude Is Constant

Let the unit vector $\hat{t}\left(t^k\right)$ represent the direction at any point of a vector field $\vec{u}$. The vector $\bar{u}$ of constant magnitude (need not necessarily be of constant magnitude) is said to be parallel along a curve C with respect to a Riemannian V_n, if its intrinsic derivative (or derived vector) in the direction of the curve at all points of C vanishes; mathematically,

$$u^i_{,k}t^k = 0 \text{ or } u^i_{,k}\frac{dx^k}{ds} = 0.$$

This can be thrown to the form:

$$\left(\frac{\partial u^i}{\partial x^k} + u^\alpha \Gamma^i_{\alpha k}\right)\frac{dx^k}{ds} = 0$$

or

$$\frac{\partial u^i}{\partial x^k}\frac{dx^k}{ds} + u^\alpha \Gamma^i_{\alpha k}\frac{dx^k}{ds} = 0,$$

so that

$$\frac{du^i}{ds} + u^\alpha \Gamma^i_{\alpha k}\frac{dx^k}{ds} = 0. \qquad (4.18.1)$$

Otherwise, the vector $\bar{u}$ is said to undergo parallel displacement (according to Levi-Civita) along the curve C of V_n if (4.18.1) is satisfied.

Increment of $\bar{u}$:

From (4.18.1), $\frac{du^i}{ds} + u^\alpha \Gamma^i_{\alpha k} \frac{dx^k}{ds} = 0$.

This can be said as the arc rate of change of the contravariant component u^i. This can be put to the form:

$du^i = -u^\alpha \Gamma^i_{\alpha k} dx^k$, which is called the **increment** of u^i due to the displacement dx^k.

4.18.2 Parallel Displacement When a Vector Is of Variable Magnitude

If the direction of two vectors is the same or if their corresponding components are proportional, they are said to be parallel.

Let $\vec{B}$ be a vector of variable magnitude parallel to a vector $\vec{A}$ at each point of the curve C in V_n so that

$$B^i = \lambda(s) A^i. \tag{4.18.2}$$

If $\vec{A}(A^i)$ is assumed to be a vector of constant magnitude parallel to the curve C with respect to V_n, then $\vec{B}(B^i)$ must also be parallel with respect to V_n along the curve but of variable magnitude.

As A^i undergoes parallel displacement,

$$A^i_{,j} \frac{dx^j}{ds} = 0. \tag{4.18.3}$$

Now,

$$B^i_{,j} \frac{dx^j}{ds} = \left(\lambda A^i\right)_{,j} \frac{dx^j}{ds} = \lambda_{,j} A^i \frac{dx^j}{ds} + \lambda A^i_{,j} \frac{dx^j}{ds}$$

$$= A^i \frac{\partial \lambda}{\partial x^j} \frac{dx^j}{ds} = A^i \frac{d\lambda}{ds} = \frac{B^i}{\lambda} \frac{d\lambda}{ds}, \quad \text{using}(4.18.3),$$

$$\therefore B^i_{,j} \frac{dx^j}{ds} = B^i f(s) \tag{4.18.4}$$

where $f(s) = \frac{d}{ds}\left[\log \lambda(s)\right]$.

The vector $\vec{B}(B^i)$ of variable magnitude should be expressed in the form (4.18.4), if it undergoes parallel displacement with respect to V_n along the curve C. It can be promptly concluded that the intrinsic derivative of $\vec{B}$ (L.H.S. of 4.18.4) at each point on the curve C has the same direction as that of $\vec{B}(B^i)$ (R.H.S.). Otherwise, $\vec{B}$ is parallel to itself along the curve C subject to the condition (4.18.4).

Writing $A^i = B^i \mu(s)$, we can get

$$A^i_{,j}\frac{dx^j}{ds} = \left[\mu B^i_{,j} + B^i\frac{\partial \mu}{\partial x^j}\right]\frac{dx^j}{ds}$$

$$= \left[\mu B^i f(s) + B^i\frac{d\mu}{ds}\right], \quad \text{using}(4.18.4).$$

$$= B^i\left[\frac{d\mu}{ds} + \mu f(s)\right]$$

We **can choose** $\frac{d\mu}{ds} + \mu f(s) = 0$ so that

$A^i_{,j}\frac{dx^j}{ds} = 0$, which is the condition of parallel displacement of the vector $\bar{A}(A^i)$ of constant magnitude along the curve C in V_n.

Hence, $\bar{B}(B^i)$ also undergoes $(\because A^i = \mu B^i)$ parallel displacement along the curve C in V_n.

Hence, proved.

Note: From the condition of parallel displacement (4.18.4),

$$B^i_{,j}\frac{dx^j}{ds} = B^i f(s)$$

$$B^k B^i_{,j}\frac{dx^j}{ds} = B^k B^i f(s) \quad \left(\text{multiplying by } B^k\right)$$

$$B^i B^k_{,j}\frac{dx^j}{ds} = B^i B^k f(s) \quad (k \leftrightarrow i).$$

Subtracting, $\left(B^i B^k_{,j} - B^k B^i_{,j}\right)\frac{dx^j}{ds} = 0$, (or) eliminating $f(s)$.

This is a modified form of the condition of parallelism of the vector $\bar{B}(B^i)$ of **variable magnitude.**

Theorem: If Two Vectors of Constant Magnitude Undergo Parallel Displacement along a Given Curve, They Incline at a Constant Angle

Proof: Let $\hat{a}(a^k)$ and $\hat{b}(b^i)$ be the two unit vectors of constant magnitudes, and θ be the angle between them:

$$\therefore \cos\theta = g_{ij}a^i b^j$$

$$\therefore \frac{d}{ds}(\cos\theta) = \frac{d}{ds}\left(g_{ij}a^i b^j\right) = \frac{\partial}{\partial x^k}\left(g_{ij}a^i b^j\right)\frac{dx^k}{ds}$$

$$= \left(g_{ij}a^i b^j\right)_{,k}\frac{dx^k}{ds} \quad \because g_{ij}a^i b^j \quad \text{is invariant}$$

$$= \left[g_{ij}\left(a^i_{,k}b^j\right) + g_{ij}\left(a^i b^j_{,k}\right)\right]\frac{dx^k}{ds} \quad \because g_{ij,k} = 0$$

$$= b_i\left(a^i_{,k}\frac{dx^k}{ds}\right) + a_j\left(b^j_{,k}\frac{dx^k}{ds}\right)$$

$$= b_i\left(a^i_{,k}t^k\right) + a_j\left(b^j_{,k}t^k\right).$$

If $\hat{a}$ and $\hat{b}$ undergo parallel displacements along a curve C given by the direction $t^k = \frac{dx^k}{ds}$, then

$$a^i_{,k}t^k = 0 = b^j_{,k}t^k.$$

$\therefore$ The above relation reduces to

$$-\sin\theta\frac{d\theta}{ds} = 0$$

$$\therefore \frac{d\theta}{ds} = 0$$

$$\sin\theta \neq 0.$$

Hence, the vectors $\hat{a}$ and $\hat{b}$ incline at a constant angle.

Hence, proved.

Theorem: If a Vector $\bar{u}$ Undergoes Parallel Displacement along a Given Curve, It Must Be of Constant Magnitude

Proof: By definition, $u^2 = g_{ij}u^i u^j = u_j u^j$

$$\therefore \frac{d}{ds}\left(u^2\right) = \frac{d}{ds}\left(u_j u^j\right) = \frac{\partial}{\partial x^k}\left(u_j u^j\right)\frac{dx^k}{ds}$$

$\because u_j u^j$ is invariant

$$= u^j\left(u_{j,k}\frac{dx^k}{ds}\right) + u_j\left(u^j_{,k}\frac{dx^k}{ds}\right)$$

$$\Rightarrow \frac{d}{ds}\left(u^2\right) = 0.$$

∴ For parallel displacement of $\bar{u}$ along the curve C of direction,

$$\frac{dx^k}{ds} = t^k,\ u_{j,k}\frac{dx^k}{ds} = 0 = u^j_{,k}\frac{dx^k}{ds}.$$

u^2 = constant, i.e., u = constant.

Hence, proved.

Exercises

1. If $\hat{a}$ and $\hat{b}$ are the two unit vectors, ϕ is a scalar invariant and the derivative of ϕ in the direction of $\hat{a}$ is $\phi_{,i}\,a^{i}$, then show that the derivative of this quantity in the direction of $\hat{b}$ is $(\phi_{i}\,a^{i})_{,j}\,b^{j} = (\phi_{,i}\,a^{i}_{,j} + a^{i}_{,ij})b^{j}$.
2. Find the nonvanishing Christoffel symbols for the metrics:
   i. $ds^2 = a^2dr^2 + \sin^2\theta d\theta^2$.
   ii. $ds^2 = dr^2 + r^2d\theta^2 + dz^2$.
   iii. $ds^2 = dr^2 + r^2d\theta^2 + r^2\sin^2\theta\, d\phi^2$.
   iv. $ds^2 = e^{-2kt}(dx^2 + dy^2 + dz^2) - dt^2$.
3. If ϕ is a scalar and $f(\phi)$ is a function of ϕ, show that

$$\nabla^2\phi = f''(\phi)(\nabla\phi)^2 + f'(\phi)\nabla^2\phi.$$

5

Properties of Curves in V_n and Geodesics

Definition

Curvature: The arc rate at which the tangent to a curve at a point P changes the direction as P moves along the curve is called the curvature (κ) of the curve.

5.1 The First Curvature of a Curve

If the coordinates x^i of the current point P of a curve C in a V_n is taken as function of the parameter arc length s, then the unit tangent vector $\hat{t}(t^i)$ to the curve is defined as $t^i = \frac{dx^i}{ds}$. The first curvature vector $\vec{P}(p^i)$ of the curve C relative to V_n is the derived vector of $\hat{t}$ along the curve which is defined as

$$p^i = t^i_{,k}\frac{dx^k}{ds}. \tag{5.1.1}$$

The first curvature vector is of magnitude $\kappa = \sqrt{g_{ij}p^ip^j}$.

From the definition (5.1.1),

$$\begin{aligned} p^i &= \left(\frac{\partial t^i}{\partial x^k} + \Gamma^i_{jk}\frac{dx^j}{ds}\right)\frac{dx^k}{ds} = \frac{\partial t^i}{\partial x^k}\frac{dx^k}{ds} + \Gamma^i_{jk}\frac{dx^j}{ds}\frac{dx^k}{ds} \\ &= \frac{dt^i}{ds} + \Gamma^i_{jk}\frac{dx^j}{ds}\frac{dx^k}{ds} = \frac{d^2x^i}{ds^2} + \Gamma^i_{jk}\frac{dx^j}{ds}\frac{dx^k}{ds}. \end{aligned} \tag{5.1.2}$$

If the first curvature of the curve p^i vanishes, then $\left(p^i =\right)\frac{d^2x^i}{ds^2} + \Gamma^i_{jk}\frac{dx^j}{ds}\frac{dx^k}{ds} = 0$ will represent a particular class of curves called **geodesics**, which will be discussed in Section 5.2.

Principal normal: The direction of the first curvature vector $\vec{P}(p^i)$ is called the principal normal, and the unit vector $\hat{n}$ in the direction of $\bar{p}$ is called the principal unit normal so that $\vec{P} = \kappa\hat{n}$.

5.2 Geodesics

Definition

A geodesic on a surface in V_n is a curve or path of extremum (or stationary) length joining any two given points on it.

For example,

i. A line joining two points in a plane is the shortest distance called straight line which is geodesic.
ii. The great circular arc joining two points on the surface of a sphere is the shortest distance between two points. Therefore, it is geodesic in three-dimensional spherical surface.

5.3 Derivation of Differential Equations of Geodesics

Let A and B be the two points on a surface in an n-dimensional space V_n. The arc length joining the two points on the surface of V_n is given by $ds^2 = g_{ij}\, dx^i dx^j$.

For stationary (or extremum) length, let us consider a small variation so that

$$2ds\delta(ds) = \frac{\partial g_{ij}}{\partial x^k}\delta x^k dx^i dx^j + g_{ij}\delta(dx^i)dx^j + g_{ij}dx^i\delta(dx^j)$$

$$= \frac{\partial g_{ij}}{\partial x^k}\delta x^k dx^i dx^j + g_{ij}d(\delta x^i)dx^j + g_{ij}dx^i d(\delta x^j) \quad i \leftrightarrow j$$

$$= \frac{\partial g_{ij}}{\partial x^k}\delta x^k dx^i dx^j + 2g_{ij}d(\delta x^i)dx^j \quad \therefore g_{ij} = g_{jt}$$

$$\therefore \delta(ds) = \left[\frac{1}{2}\frac{\partial g_{ij}}{\partial x^k}\delta x^k \frac{dx^i}{ds}\frac{dx^j}{ds} + g_{ij}\frac{d}{ds}(\delta x^i)\frac{dx^j}{ds}\right]ds.$$

Applying variational principle for extremum values of the arc length joining the two arbitrarily chosen points A and B on V_n, we are to make use of $\int_A^B \delta(ds) = 0$.

$$\therefore \int_A^B \left[\frac{1}{2}\frac{\partial g_{ij}}{\partial x^k}\delta x^k \frac{dx^i}{ds}\frac{dx^j}{ds} + g_{ij}\frac{d}{ds}(\delta x^i)\frac{dx^j}{ds}\right]ds = 0 \tag{5.3.1}$$

Let us consider the second term of (5.3.1):

$$\int_A^B \left\{ g_{ij} \frac{d}{ds}(\delta x^i) \frac{dx^j}{ds} \right\} ds = \left[g_{ij} \frac{dx^j}{ds} \cdot \delta x^i \right]_A^B - \int_A^B \left[\frac{d}{ds}\left(g_{ij} \frac{dx^j}{ds} \right) \cdot \delta x^i \right] ds$$

$$= -\int_A^B \left[\frac{d}{ds}\left(g_{ij} \frac{dx^j}{ds} \right) \delta x^i \right] ds.$$

$\because$ For small variation, $\delta x^i = 0$ at both the ends A and B.

$\therefore$ Equation (5.3.1) reduces to

$$\int_A^B \left[\frac{1}{2} \frac{\partial g_{ij}}{\partial x^k} \frac{dx^i}{ds} \frac{dx^j}{ds} \underbrace{\delta x^k} - \frac{d}{ds}\left(g_{ij} \underset{i \to k}{\frac{dx^j}{ds}} \right) \underbrace{\delta x^i} \right] ds = 0$$

or

$$\int_A^B \left[\frac{1}{2} \frac{\partial g_{ij}}{\partial x^k} \frac{dx^i}{ds} \frac{dx^j}{ds} - \left\{ \frac{\partial g_{jk}}{\partial x^i} \frac{dx^i}{ds} \frac{dx^j}{ds} + g_{kj} \frac{d^2 x^j}{ds^2} \right\} \right] \delta x^k ds = 0.$$

But for all δx^k, i.e., arbitrary value of δx^k if the above result is to hold good, then we must have

$$\frac{1}{2} \frac{\partial g_{ij}}{\partial x^k} \frac{dx^i}{ds} \frac{dx^j}{ds} - 1 \cdot \frac{\partial g_{jk}}{\partial x^i} \frac{dx^i}{ds} \frac{dx^j}{ds} - g_{jk} \frac{d^2 x^j}{ds^2} = 0$$

or

$$g_{jk} \frac{d^2 x^j}{ds^2} + \left\{ \frac{1}{2} \frac{\partial g_{jk}}{\partial x^i} \frac{dx^i}{ds} \frac{dx^j}{ds} + \frac{1}{2} \frac{d g_{ik}}{d x^j} \frac{dx^j}{ds} \frac{dx^i}{ds} \right\} - \frac{1}{2} \frac{\partial g_{ij}}{\partial x^k} \frac{dx^i}{ds} \frac{dx^j}{ds} = 0$$

(interchanging i and j in one term within the bracket).

Considering the inner product of it with g^{mk} and summing over k, we get

$$g^{mk} g_{jk} \frac{d^2 x^j}{ds^2} + g^{mk} \left\{ \Gamma_{ij,k} \right\} \frac{dx^i}{ds} \frac{dx^j}{ds} = 0$$

or

$$\delta_j^m \frac{d^2x^j}{ds^2} + \Gamma_{ij}^m \frac{dx^i}{ds}\frac{dx^j}{ds} = 0 \quad g^{mk} g_{jk} = \delta_j^m$$

or

$$\frac{d^2x^m}{ds^2} + \Gamma_{ij}^m \frac{dx^i}{ds}\frac{dx^j}{ds} = 0$$

$$\because \delta_j^m = 1 \text{ for } m = j$$

$$= 0 \text{ for } m \neq j.$$

This can be written as

$$\frac{d^2x^i}{ds^2} + \Gamma_{jk}^i \frac{dx^j}{ds}\frac{dx^k}{ds} = 0 \tag{5.3.2}$$

(changing $m \to i$ and $i \to k$).

The preceding equation is the differential equation of the curve called **geodesis.**

These are the second-order differential equations in terms of n variables x^i ($i = 1, 2, \ldots, n$); each solution must have got two arbitrary constants. Therefore, second-order n differential equations should contain $2n$ arbitrary constants in their general solutions. Hence, $2n$ given conditions are necessary to know the complete solutions.

The $2n$ coordinates of the points A and B are sufficient to determine the $2n$ arbitrary constants occurring in the solutions. Hence, the geodesic can be determined uniquely. Otherwise, through the two given points A and B on the surface in a V_n, one and only one geodesic can pass.

Again, besides the coordinates $A(x^i)$, if the n components of the unit tangent vector $\hat{t}\left(t^i = \frac{dx^i}{ds}\right)$ at the point A are known, then the geodesics can be determined uniquely.

5.4 Aliter: Differential Equations of Geodesics as Stationary Length

Let A and B be the two fixed points on a curve C of V_n, and t_0 and t_1 be the parametric values of A and B, respectively.

$\therefore$ The length of the arc joining A to B is given by

$$\int_{t_0}^{t_1} \sqrt{g_{ij}\frac{dx^i}{dt}\frac{dx^j}{dt}}\,dt$$

$$\because \left(ds^2 = g_{ij}(x^i)dx^i dx^j\right). \tag{5.4.1}$$

$$= \int_{t_0}^{t_1} \sqrt{g_{ij}\dot{x}^i\dot{x}^j}\,dt$$

If the curve C is to be geodesic, the above length should be stationary (or extremum):

But the Euler's equation for extremum value of the integral $I = \int_{t_0}^{t_1} f\left(\dot{x}^i\dot{x}^j\right)dt$ states that it must satisfy the differential equation:

$$\frac{\partial f}{\partial x^i} - \frac{d}{dt}\left(\frac{\partial f}{\partial \dot{x}^i}\right) = 0. \tag{5.4.2}$$

But for (5.4.1),

$$f = \sqrt{g_{ij}\dot{x}^i\dot{x}^j} = \frac{ds}{dt} = \dot{s}$$

$$\therefore \frac{\partial f}{\partial x^i} = \frac{1}{2\sqrt{g_{jk}\dot{x}^k\dot{x}^j}}\frac{\partial g_{jk}}{\partial x^i}\dot{x}^j\dot{x}^k = \frac{1}{2\dot{s}}\frac{\partial g_{jk}}{\partial x^i}\dot{x}^j\dot{x}^k \quad (i \to k).$$

Also $\frac{\partial f}{\partial \dot{x}^i} = \frac{1}{\dot{s}}g_{ij}\dot{x}^j$.

$\therefore$ Equation (5.4.2) reduces to

$$\frac{1}{2\dot{s}}\frac{\partial g_{jk}}{\partial x^i}\dot{x}^j\dot{x}^k - \frac{d}{dt}\left(\frac{1}{\dot{s}}g_{ij}\dot{x}^j\right) = 0$$

$$\frac{1}{2\dot{s}}\frac{\partial g_{jk}}{\partial x^i}\dot{x}^j\dot{x}^k - \left[-\frac{\ddot{s}}{\dot{s}^2}g_{ij}\dot{x}^j + \frac{1}{\dot{s}}\frac{\partial g_{ij}}{\partial x^k}\dot{x}^k\dot{x}^j + \frac{1}{\dot{s}}g_{ij}\ddot{x}^j\right] = 0$$

$$g_{ij}\ddot{x}^j + \frac{\partial g_{ij}}{\partial x^k}\dot{x}^j\dot{x}^k - \frac{1}{2}\frac{\partial g_{jk}}{\partial x^i}\dot{x}^j\dot{x}^k - \frac{\ddot{s}}{\dot{s}}g_{ij}\dot{x}^j = 0$$

$$g_{ij}\ddot{x}^j + \left(\frac{1}{2}\frac{\partial g_{ij}}{\partial x^k}\dot{x}^j\dot{x}^k + \frac{1}{2}\frac{\partial g_{ik}}{\partial x^j}\dot{x}^k\dot{x}^j\right) - \frac{1}{2}\frac{\partial g_{jk}}{\partial x^i}\dot{x}^j\dot{x}^k - \frac{\ddot{s}}{\dot{s}}g_{ij}\dot{x}^j = 0.$$

$$g_{ij}\ddot{x}^j + \Gamma_{kj,i}\,\dot{x}^j\dot{x}^k - \frac{\ddot{s}}{\dot{s}}g_{ij}\dot{x}^j = 0$$

$$g^{i\alpha}g_{ij}\ddot{x}^j + g^{i\alpha}\Gamma_{kj,i}\,\dot{x}^j\dot{x}^k - g_{ij}g^{i\alpha}\dot{x}^j\frac{\ddot{s}}{\dot{s}} = 0 \quad \left(\text{taking the inner product with } g^{i\alpha}\right)$$

$$\ddot{x}^\alpha + \Gamma_{jk}^{\alpha}\,\dot{x}^j\dot{x}^k - \dot{x}^\alpha\frac{\ddot{s}}{\dot{s}} = 0$$

If the usual arc length parameter s is chosen instead of t, then the equation transforms to

$$\therefore \frac{d^2x^i}{ds^2} + \Gamma_{jk}^{i}\frac{dx^j}{ds}\frac{dx^k}{ds} = 0$$

$$(\because \dot{s} = 1, \ddot{s} = 0).$$

This is the differential equation of geodesics obtained from the **notion of stationary** arc length.

5.5 Geodesic Is an Autoparallel Curve

Let $\hat{t}$ be the unit tangent vector to a geodesic curve. The differential equations of geodesics are

$$\frac{d^2x^i}{ds^2} + \Gamma_{jk}^{i}\frac{dx^j}{ds}\frac{dx^k}{ds} = 0$$

or

$$\frac{d}{ds}\left(\frac{dx^i}{ds}\right) + \Gamma_{jk}^{i}\frac{dx^j}{ds}\frac{dx^k}{ds} = 0$$

$$\therefore \frac{dt^i}{ds} + \Gamma_{jk}^{i}t^jt^k = 0 \quad t^i = \frac{dx^i}{ds}.$$

It can be written as

$$\frac{\partial t^i}{\partial x^k}\frac{dx^k}{ds} + \Gamma_{jk}^{i}t^jt^k = 0$$

$$\left(\frac{\partial t^i}{\partial x^k} + t^j\Gamma_{jk}^{i}\right)t^k = 0 \quad \therefore t^i_{,k}\,t^k = 0,$$

which is the condition of parallel displacement of the unit tangent vector t^i to the curve geodesic but in the direction of itself.

Hence, geodesics are autoparallel curves.

Example 1

Determine the differential equations of geodesis in a space given by the metric

$$ds^2 = -e^{2kt}\left(dx^2 + dy^2 + dz^2\right) + dt^2.$$

For the given metric, $g_{11} = g_{22} = g_{33} = -e^{2kt}$, $g_{44} = 1$.

But $g_{ij} = 0$ for $i \neq j$

$$\therefore \frac{\partial g_{11}}{\partial x^4} = \frac{\partial g_{22}}{\partial x^4} = \frac{\partial g_{33}}{\partial x^4} = -2ke^{2kt} \qquad \text{(i)}$$

$$\text{and } \frac{\partial g_{44}}{\partial x^i} = 0 \text{ for all "}i\text{"}.$$

Also for the metric, g_{ii} exists only; therefore, $g^{ii} = \frac{1}{g_{ii}}$.

Now the differential equations of geodesics are

$$\frac{d^2x^i}{ds^2} + \Gamma^i_{jk}\frac{dx^j}{ds}\frac{dx^k}{ds} = 0. \qquad \text{(ii)}$$

$\therefore$ We are to determine nonvanishing Christoffel symbols Γ^i_{jk} of the second kind:

1. $\Gamma^i_{jj} = g^{ii}\Gamma_{jj,i} = -\frac{1}{2g_{ii}}\frac{\partial g_{jj}}{\partial x^i} = -\frac{1}{2.1}\left(-2ke^{2kt}\right) = ke^{2kt}$

   due to $i = 4$ only and $j = 1, 2, 3 (i \neq j)$

   $\therefore \Gamma^4_{11} = \Gamma^4_{22} = \Gamma^4_{33} = ke^{2kt}$

2. $\Gamma^i_{ij} = g^{ii}\Gamma_{ij,i} = \frac{1}{2g_{ii}}\frac{\partial g_{ii}}{\partial x^j}$.

   $\therefore \Gamma^1_{14}, \Gamma^2_{24}, \Gamma^3_{34}$ exists only due to (i) for $i = 1, 2, 3$, and $j = 4$ only.

$$\therefore \Gamma^1_{14} = \frac{1}{2g_{11}}\frac{\partial g_{11}}{\partial x^4} = \frac{1}{2\left(-e^{2kt}\right)} \times \left(-2ke^{2kt}\right) = k.$$

   Similarly, $\Gamma^2_{24} = \Gamma^3_{34} = k$.

3. But $\Gamma^i_{ii} = \frac{1}{2g_{ii}}\frac{\partial g_{ii}}{\partial x^i} = 0 \quad \therefore \frac{\partial g_{44}}{\partial x^4} = 0$.

Now for $i = 1$, the differential equation (2) of the geodesic takes the form:

$$\frac{d^2x^1}{ds^2} + \Gamma^1_{jk} \frac{dg^j}{ds} \frac{dg^k}{ds} = 0$$

or

$$\frac{d^2x^1}{ds^2} + \Gamma^1_{14} \frac{dx^1}{ds} \frac{dx^4}{ds} + \Gamma^1_{41} \frac{dx^4}{ds} \frac{dx^1}{ds} = 0$$

or

$$\frac{d^2x}{ds^2} + 2k \frac{dx}{ds} \frac{dt}{ds} = 0 \quad \therefore \Gamma^1_{14} = \Gamma^1_{41} = k$$

or

$$\frac{\frac{d^2x}{ds^2}}{\frac{dx}{ds}} = -2k \frac{dt}{ds}.$$

Integrating, $\log\left(\frac{dx}{ds}\right) = -2kt + \log a$.

This can be written as

$$\frac{dx}{ds} = ae^{-2kt}. \qquad \text{(iii)}$$

Similarly, putting $i = 2$ and 3, respectively, we can get

$$\frac{dy}{ds} = +be^{-2kt}. \qquad \text{(iv)}$$

$$\frac{dz}{ds} = ce^{-2kt}. \qquad \text{(v)}$$

Again putting $i = 4$, Equation (ii) can be simplified as

$$\frac{d^2x^4}{ds^2} + \Gamma^4_{jk} \frac{dx^j}{ds} \frac{dx^k}{ds} = 0$$

or

$$\frac{d^2x^4}{ds^2} + \Gamma^4_{11} \frac{dx^1}{ds} \frac{dx^1}{ds} + \Gamma^4_{22} \frac{dx^2}{ds} \frac{dx^2}{ds} + \Gamma^4_{33} \frac{dx^3}{ds} \frac{dx^3}{ds} = 0$$

or

$$\frac{d^2t}{ds^2} + ke^{2kt} \left[\left(\frac{dx}{ds}\right)^2 + \left(\frac{dy}{ds}\right)^2 + \left(\frac{dz}{ds}\right)^2 \right] = 0$$

or

$$\frac{d^2t}{ds^2} + ke^{2kt} \times \left(a^2 + b^2 + c^2\right)e^{-4kt} = 0 \text{ using (iii), (iv), and (v)}$$

$$\therefore \frac{d^2t}{ds^2} + k\alpha^2 e^{-2kt} = 0 \qquad \text{(vi)}$$

putting $\alpha^2 = \left(a^2 + b^2 + c^2\right)$.

Hence, Equations (iii)–(vi) are the required differential equations for the geodesics.

5.6 Integral Curve of Geodesic Equations

The equations of geodesics are (Equation 5.3.2)

$$\frac{d^2x^i}{ds^2} + \Gamma^i_{jk}\frac{dx^j}{ds}\frac{dx^k}{ds} = 0. \qquad (5.6.1)$$

If C is the curve geodesic through a point P and s is the arc length, then by Taylor's theorem, we can write

$$x^i = x^i_0 + \left(\frac{dx^i}{ds}\right)_0 s + \frac{1}{\underline{|2}}\left(\frac{d^2x^i}{ds^2}\right)_0 s^2 + \frac{1}{\underline{|3}}\left(\frac{d^3x^i}{ds^3}\right)_0 s^3 + \cdots. \qquad (5.6.2)$$

The second and higher derivatives of x^i with respect to s can be determined from the equations of geodesics (5.6.1) as follows:

$$\frac{d^2x^i}{ds^2} = -\Gamma^i_{jk}\frac{dx^j}{ds}\frac{dx^k}{ds} \qquad (5.6.3)$$

$$\frac{d^3x^i}{ds^3} = -\left[\frac{d}{ds}\left(\Gamma^i_{jk}\right)\frac{dx^j}{ds}\frac{dx^k}{ds} + \Gamma^i_{jk}\frac{d^2x^j}{ds^2}\frac{dx^k}{ds} + \Gamma^i_{jk}\frac{dx^j}{ds}\frac{d^2x^k}{ds^2}\right]$$

$$= -\left[\frac{\partial}{\partial x^l}\left(\Gamma^i_{jk}\right)\frac{dx^j}{ds}\frac{dx^k}{ds}\frac{dx^l}{ds} - \Gamma^i_{jk}\Gamma^j_{\alpha\beta}\underset{(\alpha\to l,\ \beta\to j,\ j\to\alpha)}{\frac{dx^\alpha}{ds}\frac{dx^\beta}{ds}\frac{dx^k}{ds}} - \Gamma^i_{jk}\Gamma^k_{\alpha\beta}\underset{(\alpha\to l,\ \beta\to k,\ k\to\alpha)}{\frac{dx^\alpha}{ds}\frac{dx^\beta}{ds}\frac{dx^j}{ds}}\right]$$

using (5.6.3)

$$= -\left[\frac{\partial}{\partial x^l}\left(\Gamma^i_{jk}\right)\underbrace{\frac{dx^j}{ds}\frac{dx^k}{ds}\frac{dx^l}{ds}} - \Gamma^i_{\alpha k}\underset{(j\leftrightarrow k)}{\Gamma^\alpha_{lj}}\underbrace{\frac{dx^j}{ds}\frac{dx^k}{ds}\frac{dx^l}{ds}} - \Gamma^i_{j\alpha}\Gamma^\alpha_{lk}\underbrace{\frac{dx^j}{ds}\frac{dx^k}{ds}\frac{dx^l}{ds}}\right]$$

$$\therefore \frac{d^3x^i}{ds^3} + \Gamma^i_{jkl}\frac{dx^j}{ds}\frac{dx^k}{ds}\frac{dx^l}{ds} = 0, \tag{5.6.4}$$

where

$$\Gamma^i_{jkl} = \frac{1}{3}P\left[\frac{\partial}{\partial x^l}\left(\Gamma^i_{jk}\right) - \Gamma^i_{\alpha j}\Gamma^\alpha_{lk} - \Gamma^i_{j\alpha}\Gamma^\alpha_{lk}\right]$$

$$= \frac{1}{3}P\left[\frac{\partial}{\partial x^l}\left(\Gamma^i_{jk}\right) - 2\Gamma^i_{\alpha j}\Gamma^\alpha_{kl}\right]$$

and P indicates the sum of the terms obtained by permuting the subscripts cyclicly.*

Using (5.6.3) and (5.6.4) in (5.6.2) for $P(x_o)$, we get

$$x^i = x^i_o + \left(\frac{dx^i}{ds}\right)_o s - \frac{1}{\underline{|2}}\left(\Gamma^i_{jk}\right)_o\left(\frac{dx^j}{ds}\right)_o\left(\frac{dx^k}{ds}\right)s^2 - \frac{1}{\underline{|3}}\left(\Gamma^i_{jkl}\right)_o$$

$$\left(\frac{dx^j}{ds}\right)_o\left(\frac{dx^k}{ds}\right)_o\left(\frac{dx^l}{ds}\right)_o s^3 + \cdots.$$

If we define $\xi^i = \left(\frac{dx^i}{ds}\right)_o$, it can be thrown to the form:

$$x^i = x^i_o + \xi^i s - \frac{1}{\underline{|2}}\left(\Gamma^i_{jk}\right)_o \xi^j\xi^k s^2 - \frac{1}{\underline{|3}}\left(\Gamma^i_{jkl}\right)_o \xi^j\xi^k\xi^l s^3 + \cdots \tag{5.6.5}$$

The convergence of the series is dependent on ξ^i and g_{ij} (*for* Γ^i_{jk}). This, of course, represents an **integral (solution) curve** of the equations of geodesics (5.6.1) for small values of the parameter s.

5.7 Riemannian and Geodesic Coordinates, and Conditions for Riemannian and Geodesic Coordinates

i. **Riemannian coordinates:** Consider $y^i = \xi^i s$ for the particular geodesic* (5.6.5) which passes through the point $P(x_o)$. It takes the form:

$$x^i = x^i_o + y^i - \frac{1}{\underline{|2}}\left(\Gamma^i_{jk}\right)_o y^j y^k - \frac{1}{\underline{|3}}\left(\Gamma^i_{jkl}\right)_o y^j y^k y^l + \cdots. \tag{5.7.1}$$

* [2, p. 52].

It represents all geodesics passing through $P_o(x_o)$ given by various directions $\xi^j = \left(\frac{dx^i}{ds}\right)_o$. Also the equations $y^i = \xi^i s$ define a curve otherwise geodesic for the given set of values of ξ^i in terms of new coordinates y^i. These coordinates y^i's are called **Riemannian coordinates** as adopted by Riemann. Of course, these coordinates have got their own domain about P_o.

Let the fundamental form, Christoffel symbols in terms of the Riemannian coordinates y^i, be represented by*

$$\bar{g}_{ij}(y^i)dy^i dy^j, \quad \overline{\Gamma}_{ij,k} \text{ and } \overline{\Gamma}^i_{jk}, \text{ respectively.}$$

∴ The equations of geodesics are

$$\frac{d^2y^i}{ds^2} + \overline{\Gamma}^i_{jk}\frac{dy^j}{ds}\frac{dy^k}{ds} = 0 \tag{5.7.2}$$

$$\therefore \overline{\Gamma}^i_{jk}\xi^j\xi^k = 0 \quad \because y^i = \xi^i s$$

$$\therefore \overline{\Gamma}^i_{jk}y^j y^k = 0$$

Clearly, subject to these conditions, the above equations of geodesics are satisfied, and hence, y^i's are Riemannian coordinates.

Following the method adopted in Section 5.6, but for Equation (5.7.2) in terms of the Riemannian coordinates y^i, we can get

$$y^i = \xi^i s - \frac{1}{2}\left(\overline{\Gamma}^i_{jk}\right)_o \xi^j\xi^k s^2 + \cdots, (y^i)_o = 0 \quad \text{for } s = 0.$$

This gives $\left(\overline{\Gamma}^i_{jk}\right)_o = 0$ if it is to reduce to $y^i = \xi^i s$.

$$\therefore \left(\overline{\Gamma}_{ij,k}\right) = 0 \Rightarrow \left(\frac{\partial \bar{g}_{ij}}{\partial y^k}\right)_o = 0 \quad (i, j, k = 1, 2, \ldots, n)$$

$$\bar{g}_{ij,k} = \left(\frac{\partial \bar{g}_{ij}}{\partial y^k}\right)_o - \left[\bar{g}_{\alpha j}\overline{\Gamma}^\alpha_{ik} + \bar{g}_{\alpha i}\overline{\Gamma}^\alpha_{jk}\right]_o = \left(\frac{\partial \bar{g}_{ij}}{\partial y^k}\right)_o = 0.$$

∴The first covariant derivative of the components of the fundamental tensor in these coordinates at the **origin** must vanish. This is the condition of y^i's to be Riemannian coordinates.

ii. Geodesic coordinates:

In an arbitrary Riemannian V_n, it is not possible to choose the Cartesian coordinate system, where the metric function g_{ij}'s are constants. But it is possible to choose the coordinate system where g_{ij}'s are locally constants so that

$$\frac{\partial g_{ij}}{\partial x^k} = 0 \quad \text{at the point called the pole,}$$
$$\neq 0 \quad \text{elsewhere.}$$

$\therefore \Gamma^i_{jk} = 0$ at the pole P_o; such a system of coordinates are called **geodesic coordinates** with pole P_o. Hence, for any tensor $A_{i,j} = \dfrac{\partial A_i}{\partial x^j}$ which is a condition for geodesic coordinates.

5.7.1 Another Form of Condition for Geodesic Coordinates

From the transformation law (Section 4.4), the resulting equation (4.4.5) of Christoffel symbol of the second kind, we have

$$\Gamma'^p_{ij}\frac{\partial x^\alpha}{\partial x'^p} = \Gamma^\alpha_{\lambda\mu}\frac{\partial x^\lambda}{\partial x'^i}\frac{\partial x^\mu}{\partial x'^j} + \frac{\partial^2 x^\alpha}{\partial x'^i \partial x'^j}$$

$$\therefore -\Gamma'^\alpha_{\lambda\mu}\frac{\partial x'^\lambda}{\partial x^i}\frac{\partial x'^\mu}{\partial x^j} = \frac{\partial^2 x'^\alpha}{\partial x^i \partial x^j} - \Gamma^p_{ij}\frac{\partial x'^\alpha}{\partial x^p} \left(\text{interchanging } x^i \text{ and } x'^i \text{ systems}\right)$$

$$= \frac{\partial}{\partial x^i}\left(\frac{\partial x'^\alpha}{\partial x^j}\right) - \Gamma^p_{ji}\left(\frac{\partial x'^\alpha}{\partial x^p}\right) \tag{5.7.3}$$

$$= \left(\frac{\partial x'^\alpha}{\partial x^j}\right)_{,i}.$$

Taking a **fixed** value of x'^α out of n independent values, we can entrust $\dfrac{\partial x'^\alpha}{\partial x^j} = x'^\alpha_{,j}$ $\because x'^\alpha$ is a scalar invariant for the fixed value.

$\therefore$ Equation (5.7.3) can be written as

$$-\Gamma'^\alpha_{\lambda\mu}\frac{\partial x'^\lambda}{\partial x^i}\frac{\partial x'^\mu}{\partial x^j} = x'^\alpha_{,ij}, \tag{5.7.4}$$

i.e., a second covariant derivative with respect to the metric of V_n. But if x'^α are geodesic coordinates, then $\Gamma'^\alpha_{\lambda\mu} = 0$.

$$\therefore x'^\alpha_{,ij} = 0.$$

$\therefore$ The second covariant derivatives of the geodesic coordinates x'^{α} must vanish.

Otherwise, if $x'^{\alpha}_{,ij} = 0$, then from (5.7.4), $\Gamma'^{\alpha}_{\lambda\mu} = 0$ at pole.

Hence, if a system of coordinates are geodesic coordinates with pole, then their second covariant derivatives with respect to the metric of the space must vanish at that point. From the above interpretation, it can be concluded that it is a necessary and sufficient condition.

5.8 If a Curve Is a Geodesic of a Space (V_m), It Is also a Geodesic of Any Space V_n in Which It Lies (V_n a Subspace)

Let $C(x^i)$ be any non-minimal curve in a V_n at points $x^i(s)$ with fundamental form $ds^2 = g_{ij}\, dx^i\, dx^j$. If $\lambda^i = \dfrac{dx^i}{ds}$ are the components of a **unit vector field** of V_n, then the derived vector of λ^i in the direction of C is given by

$$\lambda^i_{,k} \frac{dx^k}{ds} = \eta^i \text{(say)}. \tag{5.8.1}$$

But at points of C, if the vectors are parallel in the direction of C, then $\eta^i = \lambda^i_{,k} \dfrac{dx^k}{ds} = 0.$

Let the space V_n be **immersed** in a space V_m of coordinates y^{α} and fundamental form $ds^2 = a_{\alpha\beta} dy^{\alpha} dy^{\beta}$ so that $y^{\alpha} = f^{\alpha}(x^i) \quad (i = 1,2,\ldots,n; \quad \alpha = 1,2,\ldots,m)$

$$\therefore a_{\alpha\beta} = g_{ij} \frac{\partial x^i}{\partial y^{\alpha}} \frac{\partial x^j}{\partial y^{\beta}} \text{ or } \quad g_{ij} = a_{\alpha\beta} \frac{\partial y^{\alpha}}{\partial x^i} \frac{\partial y^{\beta}}{\partial x^j}. \tag{5.8.2}$$

Let the components in terms of y's of the vector field in V_m be ξ^{α}, and λ^i's be the corresponding components in x's.

$$\therefore \frac{dy^{\alpha}}{ds} = \frac{\partial y^{\alpha}}{\partial x^i} \frac{dx^i}{ds} \quad \therefore \xi^{\alpha} = \frac{\partial y^{\alpha}}{\partial x^i} \lambda^i$$

$$\therefore \frac{d\xi^{\alpha}}{ds} = \frac{d\lambda^i}{ds} \frac{\partial y^{\alpha}}{\partial x^i} + \lambda^i \frac{\partial^2 y^{\alpha}}{\partial x^i \partial x^j} \frac{dx^j}{ds} \tag{5.8.3}$$

$$\therefore \xi^{\beta}_{,\alpha} \frac{dy^{\alpha}}{ds^i} = \eta^{\beta} = \left[\frac{\partial \xi^{\beta}}{\partial y^{\alpha}} + \xi^{\gamma} \left(\Gamma^{\beta}_{\gamma\alpha} \right)_a \right] \frac{dy^{\alpha}}{ds},$$

where $\left(\Gamma_{\gamma\alpha}^{\beta}\right)_a$ denotes the Christoffel (bracket) symbol of the second kind with respect to the fundamental tensor $a_{\alpha\beta}$.

$$\therefore \eta^{\beta} = \frac{\partial \xi^{\beta}}{\partial y^{\alpha}}\frac{dy^{\alpha}}{ds} + \left(\Gamma_{\gamma\alpha}^{\beta}\right)_a \xi^{\gamma}\frac{dy^{\alpha}}{ds} = \frac{d\xi^{\beta}}{ds} + \left(\Gamma_{\gamma\alpha}^{\beta}\right)_a \lambda^{i}\frac{\partial y^{\gamma}}{\partial x^{i}}\frac{\partial y^{\alpha}}{\partial x^{j}}\frac{dx^{j}}{ds}, \quad \text{using (5.8.3)}$$

$$\therefore \eta^{\beta} = \frac{d\lambda^{i}}{ds}\underset{(i\to j)}{\frac{\partial y^{\beta}}{\partial x^{i}}} + \lambda^{j}\frac{\partial^2 y^{\beta}}{\partial x^{i}\partial x^{j}}\frac{dx^{i}}{ds} + \left(\Gamma_{\gamma\alpha}^{\beta}\right)_a \lambda^{i}\frac{\partial y^{\gamma}}{\partial x^{i}}\frac{\partial y^{\alpha}}{\partial x^{j}}\underset{(i\leftrightarrow j)}{\frac{dx^{j}}{ds}}. \tag{5.8.4}$$

Now,

$$\left(\Gamma_{ij,k}\right)_g = \frac{1}{2}\left[\frac{\partial g_{ik}}{\partial x^{j}} + \frac{\partial g_{jk}}{\partial x^{i}} - \frac{\partial g_{ij}}{\partial x^{k}}\right]$$

$$= \frac{1}{2}\left[\frac{\partial a_{\alpha\gamma}}{\partial y^{\beta}}\frac{\partial y^{\beta}}{\partial x^{j}}\frac{\partial y^{\alpha}}{\partial x^{i}}\frac{\partial y^{\gamma}}{\partial x^{k}} + a_{\alpha\gamma}\frac{\partial^2 y^{\alpha}}{\partial x^{i}\partial x^{j}}\frac{\partial y^{\gamma}}{\partial x^{k}} \quad \because g_{ik} = a_{\alpha\gamma}\frac{\partial y^{\alpha}}{\partial x^{i}}\frac{\partial y^{\gamma}}{\partial x^{k}}, \text{ etc.}\right.$$

$$+ a_{\alpha\gamma}\frac{\partial y^{\alpha}}{\partial x^{i}}\frac{\partial^2 y^{\gamma}}{\partial x^{j}\partial x^{k}} + \frac{\partial a_{\beta\gamma}}{\partial y^{\alpha}}\frac{\partial y^{\alpha}}{\partial x^{i}}\frac{\partial y^{\beta}}{\partial x^{j}}\frac{\partial y^{\gamma}}{\partial x^{k}} + a_{\beta\gamma}\underset{(\beta\to\alpha)}{\frac{\partial^2 y^{\beta}}{\partial x^{i}\partial x^{j}}}\frac{\partial y^{\gamma}}{\partial x^{k}}$$

$$\begin{Bmatrix} i & j & k \\ \alpha & \beta & \gamma \end{Bmatrix}$$

$$\left.- \frac{\partial a_{\alpha\beta}}{\partial y^{\gamma}}\cdot\frac{\partial y^{\gamma}}{\partial x^{k}}\frac{\partial y^{\alpha}}{\partial x^{i}}\frac{\partial y^{\beta}}{\partial x^{j}} - a_{\alpha\beta}\underset{(\alpha\to\gamma)}{\frac{\partial^2 y^{\alpha}}{\partial x^{i}\partial x^{k}}}\frac{\partial y^{\beta}}{\partial x^{j}} - a_{\alpha\beta}\frac{\partial y^{\alpha}}{\partial x^{i}}\underset{(\beta\to\gamma)}{\frac{\partial^2 y^{\beta}}{\partial x^{j}\partial x^{k}}}\right]$$

$$= \frac{1}{2}\left[\left(\frac{\partial a_{\alpha\gamma}}{\partial y^{\beta}} + \frac{\partial a_{\beta\gamma}}{\partial y^{\alpha}} - \frac{\partial a_{\alpha\beta}}{\partial y^{\gamma}}\right)\frac{\partial y^{\alpha}}{\partial x^{i}}\frac{\partial y^{\beta}}{\partial x^{j}}\frac{\partial y^{\gamma}}{\partial x^{k}} + 2a_{\alpha\gamma}\frac{\partial^2 y^{\alpha}}{\partial x^{i}\partial x^{j}}\frac{\partial y^{\gamma}}{\partial x^{k}}\right]$$

$$\left(\Gamma_{ij,k}\right)_g = \left(\Gamma_{\alpha\beta,\gamma}\right)_a \frac{\partial y^{\alpha}}{\partial x^{i}}\frac{\partial y^{\beta}}{\partial x^{j}}\frac{\partial y^{\gamma}}{\partial x^{k}} + a_{\alpha\gamma}\underset{(\alpha\to c)}{\frac{\partial^2 y^{\alpha}}{\partial x^{i}\partial x^{j}}}\frac{\partial y^{\gamma}}{\partial x^{k}}$$

$$= \left(a_{c\gamma}\Gamma_{\alpha\beta}^{c}\right)_a \frac{\partial y^{\alpha}}{\partial x^{i}}\frac{\partial y^{\beta}}{\partial x^{j}}\frac{\partial y^{\gamma}}{\partial x^{k}} + a_{c\gamma}\frac{\partial^2 y^{c}}{\partial x^{i}\partial x^{j}}\frac{\partial y^{\gamma}}{\partial x^{k}}$$

$$= a_{c\gamma}\left[\frac{\partial^2 y^{c}}{\partial x^{i}\partial x^{j}} + \left(\Gamma_{\alpha\beta}^{c}\right)_a \frac{\partial y^{\alpha}}{\partial x^{i}}\frac{\partial y^{\beta}}{\partial x^{j}}\right]\frac{\partial y^{\gamma}}{\partial x^{k}}$$

$$\left(\Gamma_{ij,k}\right)_g = a_{\beta\gamma}\left[\frac{\partial^2 x^{\beta}}{\partial x^{i}\partial x^{j}} + \underset{(c\leftrightarrow\gamma)}{\left(\Gamma_{\alpha c}^{\beta}\right)_a} \frac{\partial y^{c}}{\partial x^{j}}\frac{\partial y^{\alpha}}{\partial x^{i}}\right]\frac{\partial y^{\gamma}}{\partial x^{k}}. \tag{5.8.5}$$

Multiplying (5.8.4) by $a_{\beta c}\dfrac{\partial y^c}{\partial x^k}$,

$$a_{\beta c}\frac{\partial y^c}{\partial x^k}\eta^\beta = \frac{d\lambda^j}{ds}a_{\beta c}\underbrace{\frac{\partial y^c}{\partial x^k}\frac{\partial y^\beta}{\partial x^j}}$$

$$+\lambda^j\frac{dx^i}{ds}a_{\beta c}\frac{\partial y^c}{\partial x^k}\left[\frac{\partial^2 x^\beta}{\partial x^i\partial x^j}+\left(\Gamma_{\alpha\gamma}^{\beta}\right)_a\frac{\partial y^\alpha}{\partial x^i}\frac{\partial y^\gamma}{\partial x^j}\right]$$

$$= g_{jk}\frac{d\lambda^j}{ds}+\lambda^j\frac{dx^i}{ds}\left(\Gamma_{ij,k}\right)_g \quad \text{using}(5.8.5)$$

$$= g_{jk}\frac{\partial\lambda^j}{\partial x^i}\frac{dx^i}{ds}+\lambda^j\frac{dx^i}{ds}\left(\Gamma_{ij,k}\right)_g$$

$$= g_{jk}\frac{dx^i}{ds}\underset{(j\to l)}{\frac{\partial\lambda^j}{\partial x^i}}+\lambda^j\frac{dx^i}{ds}\left(g_{lk}\Gamma_{ij}^{l}\right)_g$$

$$= g_{lk}\frac{dx^i}{ds}\left[\frac{\partial\lambda^l}{\partial x^i}+\lambda^j\left(\Gamma_{ij}^{l}\right)_g\right]$$

$$= g_{lk}\left(\frac{dx^i}{ds}\lambda^l_{,i}\right).$$

If ξ^β is parallel with respect to the curve C in V_m, then $\xi^\beta_{,\alpha}\dfrac{dy^\alpha}{ds}=\eta^\beta=0$; therefore, $\lambda^l_{,i}\dfrac{dx^i}{ds}=0$.

$\therefore$ The vectors that are parallel to a curve C in V_m, they are also parallel with respect to V_n, a subspace of V_m.

If λ^i is a **unit tangent** vector to the curve C, then parallelism of the vectors along C means that the **curve is a geodesic**.

Hence, if a curve is a geodesic in a space (here V_m), it is also a geodesic in the subspace V_n as it immerses in V_m.

Hence, proved.

Exercises

1. Determine the differential equations of geodesics for the metric:
   i. $ds^2=dr^2+r^2d\theta^2+r^2\sin^2\theta d\phi^2$.
   ii. $ds^2=dr^2+r^2d\theta^2+dz^2$.

2. Find the differential equations of geodesics for the metric:

$$ds^2 = f(x)dx^2 + dy^2 + dz^2 + \frac{1}{f(x)}dt^2.$$

3. Applying variational principles, derive the differential equations of geodesic:

$$\frac{d^2x^i}{ds^2} + \Gamma^i_{jk}\frac{dx^j}{ds}\frac{dx^k}{ds} = 0.$$

4. If the coordinates x^i of points on a geodesic are the functions of s, the arc length, show that

   $\frac{d^r\phi}{ds^r} = \phi_{,ij...p}\frac{dx^i}{ds}\frac{dx^j}{ds}\cdots\frac{dx^p}{ds}$, where ϕ is any scalar function of x's.

5. Obtain the differential equations of geodesic, if

   $ds^2 = \frac{dt^2}{(1-kx)} - \frac{1}{c^2(1-kx)^2}(dx^2 + dy^2 + dz^2)$, k = constant. Hence, prove that along a geodesic, $V^2 - v^2 = kc^2x$, where V is constant and $v^2 = \left(\frac{dx}{dt}\right)^2 + \left(\frac{dy}{dt}\right)^2 + \left(\frac{dz}{dt}\right)^2$.

6. Find the differential equations of geodesic for the metric:

   $ds^2 = -dx^2 - dy^2 - dz^2 + f(x, y, z)\,dt^2$.

7. Obtain the differential equations of geodesic as a means of stationary arc length.

8. Obtain the differential equations of geodesic for the metric:

$$ds^2 = dx^2 + dy^2 + dz^2 + 2gt\,dx\,dt - c^2\left(1 - \frac{g^2t^2}{c^2}\right)dt^2.$$

6

Riemann Symbols (Curvature Tensors)

6.1 Introduction

The importance of intrinsic multidimensional differential geometry has found its place in the study of general theory of relativity. The general theory of relativity is inherently related to the geometry of curved space due to the effects of gravity. For this reason, Einstein needs to deal with four-dimensional space–time continuum in support of "absolute differential calculus" or tensor calculus. The Riemannian geometry associated with covariant differentiation through the fundamental tensor g_{ij} characterizing the space is properly suited for the curved space of general theory of relativity. This demands, though not general, an important notion (or concept), namely, **Riemannian symbols** or **curvature tensors**.

6.2 Riemannian Tensors (Curvature Tensors)

The x^j-covariant differentiation of the covariant vector A_i with respect to the fundamental tensor g_{ij} is given by

$$A_{i,j} = \frac{\partial A_i}{\partial x^j} - A_\alpha \Gamma_{ij}^{\alpha}, \tag{6.2.1}$$

which is again a second-order covariant tensor. Hence, considering its x^k-covariant differentiation again, it can be written as

$$A_{i,jk} = \frac{\partial A_{i,j}}{\partial x^k} - A_{a,j}\Gamma_{ik}^{a} - A_{i,a}\Gamma_{jk}^{a} \quad \left(\text{treating } A_{ij} = A_{i,j}\right)$$

$$= \frac{\partial}{\partial x^k}\left[\frac{\partial A_i}{\partial x^j} - A_\alpha \Gamma_{ij}^{\alpha}\right] - \left(\frac{\partial A_a}{\partial x^j} - A_\alpha \Gamma_{aj}^{\alpha}\right)\Gamma_{ik}^{a} - \left(\frac{\partial A_i}{\partial x^a} - A_\alpha \Gamma_{ia}^{\alpha}\right)\Gamma_{jk}^{a} \text{ using} (6.2.1)$$

$$= \frac{\partial^2 A^i}{\partial x^k \partial x^j} - \Gamma_{ij}^{\alpha} \frac{\partial A_\alpha}{\partial x^k} - \frac{\partial}{\partial x^k}\left(\Gamma_{ij}^{\alpha}\right) A_\alpha - \frac{\partial A_a}{\partial x^j}\Gamma_{ik}^{a} + A_\alpha \Gamma_{aj}^{\alpha}\Gamma_{ik}^{a} - \frac{\partial A_i}{\partial x^a}\Gamma_{jk}^{a} + A_\alpha \Gamma_{ia}^{\alpha}\Gamma_{jk}^{a}$$

$$A_{i,jk} = \left[\frac{\partial^2 A_i}{\partial x^k \partial x^j} - \frac{\partial A_\alpha}{\partial x^k}\Gamma_{ij}^{\alpha} - \frac{\partial A_a}{\partial x^j}\Gamma_{ik}^{a} - \frac{\partial A_i}{\partial x^a}\Gamma_{jk}^{a} + A_\alpha \Gamma_{ia}^{\alpha}\Gamma_{jk}^{a}\right]$$

$$- \frac{\partial}{\partial x^k}\left(\Gamma_{ij}^{\alpha}\right) A_\alpha + A_\alpha \Gamma_{aj}^{\alpha}\Gamma_{ik}^{a}.$$

(6.2.2)

As covariant differentiation is **not commutative**, we can write (interchanging j and k)

$$A_{i,kj} = \left[\frac{\partial^2 A_i}{\partial x^j \partial x^k} - \underset{(\alpha \to a)}{\frac{\partial A_\alpha}{\partial x^j}}\Gamma_{ik}^{\alpha} - \underset{(a \to \alpha)}{\frac{\partial A_a}{\partial x^k}}\Gamma_{ij}^{a} - \frac{\partial A_i}{\partial x^a}\Gamma_{kj}^{a} + A_\alpha \Gamma_{ia}^{\alpha}\Gamma_{kj}^{a}\right] \tag{6.2.3}$$

$$- \frac{\partial}{\partial x^j}\left(\Gamma_{ik}^{\alpha}\right) A_\alpha + A_\alpha \Gamma_{ak}^{\alpha}\Gamma_{ij}^{a}.$$

Subtracting (6.2.3) from (6.2.2), we get

$$A_{i,jk} - A_{i,kj} = -A_\alpha \frac{\partial}{\partial x^k}\left(\Gamma_{ij}^{\alpha}\right) + A_\alpha \Gamma_{aj}^{\alpha}\Gamma_{ik}^{a} + \frac{\partial}{\partial x^j}\left(\Gamma_{ik}^{\alpha}\right) A_\alpha - A_\alpha \Gamma_{ak}^{\alpha}\Gamma_{ij}^{a}$$

$$= A_\alpha \left[\frac{\partial}{\partial x^j}\left(\Gamma_{ik}^{\alpha}\right) - \frac{\partial}{\partial x^k}\left(\Gamma_{ij}^{\alpha}\right) + \Gamma_{aj}^{\alpha}\Gamma_{ik}^{a} - \Gamma_{ak}^{\alpha}\Gamma_{ij}^{a}\right] \tag{6.2.4}$$

$$A_{i,jk} - A_{i,kj} = A_\alpha R_{ijk}^{\alpha},$$

where

$$R_{ijk}^{\alpha} = \frac{\partial}{\partial x^j}\left(\Gamma_{ik}^{\alpha}\right) - \frac{\partial}{\partial x^k}\left(\Gamma_{ij}^{\alpha}\right) + \Gamma_{aj}^{\alpha}\Gamma_{ik}^{a} - \Gamma_{ak}^{\alpha}\Gamma_{ij}^{a} \tag{6.2.5}$$

The left-hand side of (6.2.4) being the difference of two third-order covariant tensors must also be a third-order covariant tensor with real covariant indices, i, j, k. Hence, the right-hand side must also be a third-order covariant tensor with real indices i, j, k. But already there appears the first-order covariant tensor A_α on the right-hand side of (6.2.4). This necessitates the use of real covariant indices i, j, k in the symbol R_{ijk}^{α} with dummy index α in place of the bracket to yield a covariant tensor only.

$\therefore$ By quotient law, R_{ijk}^{α} must be a fourth-order mixed tensor with its explicit form:

$$R^{\alpha}_{ijk} = \frac{\partial}{\partial x^j}\left(\Gamma^{\alpha}_{ik}\right) - \frac{\partial}{\partial x^k}\left(\Gamma^{\alpha}_{ij}\right) + \Gamma^{\alpha}_{aj}\Gamma^{a}_{ik} - \Gamma^{\alpha}_{ak}\Gamma^{a}_{ij}.$$

This is called the **curvature tensor** for the Riemannian metric $ds^2 = g_{ij}dx^i dx^j$. Also R^{α}_{ijk} is referred to as the Riemannian symbols of the second kind.

6.3 Derivation of the Transformation Law of Riemannian Tensor R^{α}_{abc}

From the transformation law of Christoffel bracket (symbol) (4.4.5) of the second kind, we get

$$\Gamma'^{p}_{ij}\frac{\partial x^{\alpha}}{\partial x'^{p}} = \Gamma^{\alpha}_{ab}\frac{\partial x^{a}}{\partial x'^{i}}\frac{\partial x^{b}}{\partial x'^{j}} + \frac{\partial^2 x^{\alpha}}{\partial x'^{i}\partial x'^{j}}. \tag{6.3.1}$$

$$\begin{bmatrix} i & j & k \\ a & b & c \end{bmatrix}.$$

Differentiating partially with respect to x'^k,

$$\frac{\partial}{\partial x'^{k}}\left(\Gamma'^{p}_{ij}\right)\frac{\partial x^{\alpha}}{\partial x'^{p}} + \Gamma'^{p}_{ij}\frac{\partial^2 x^{\alpha}}{\partial x'^{k}\partial x'^{p}}$$

$$= \frac{\partial}{\partial x^{c}}\left(\Gamma^{\alpha}_{ab}\right)\frac{\partial x^{c}}{\partial x'^{k}}\frac{\partial x^{a}}{\partial x'^{i}}\frac{\partial x^{b}}{\partial x'^{j}} + \Gamma^{\alpha}_{ab}\frac{\partial^2 x^{a}}{\partial x'^{i}\partial x'^{k}}\frac{\partial x^{b}}{\partial x'^{j}} \tag{6.3.2}$$

$$+\Gamma^{\alpha}_{ab}\frac{\partial x^{a}}{\partial x'^{i}}\frac{\partial^2 x^{b}}{\partial x'^{j}\partial x'^{k}} + \frac{\partial^3 x^{\alpha}}{\partial x'^{i}\partial x'^{j}\partial x'^{k}}.$$

Interchanging j and k in the above expression,

$$\frac{\partial}{\partial x'^{j}}\left(\Gamma'^{p}_{ik}\right)\frac{\partial x^{\alpha}}{\partial x'^{p}} + \Gamma'^{p}_{ik}\frac{\partial^2 x^{\alpha}}{\partial x'^{j}\partial x'^{p}}$$

$$= \frac{\partial}{\partial x^{c}}\left(\underset{(c\leftrightarrow b)}{\Gamma^{\alpha}_{ab}}\right)\frac{\partial x^{c}}{\partial x'^{j}}\frac{\partial x^{a}}{\partial x'^{i}}\frac{\partial x^{b}}{\partial x'^{k}} + \Gamma^{\alpha}_{ab}\underset{(b\to c)}{\frac{\partial^2 x^{a}}{\partial x'^{i}\partial x'^{j}}}\frac{\partial x^{b}}{\partial x'^{k}} \tag{6.3.3}$$

$$+\Gamma^{\alpha}_{ab}\frac{\partial x^{a}}{\partial x'^{i}}\frac{\partial^2 x^{b}}{\partial x'^{k}\partial x'^{j}} + \frac{\partial^3 x^{\alpha}}{\partial x'^{i}\partial x'^{k}\partial x'^{j}}.$$

Eliminating $\frac{\partial^3 x^{\alpha}}{\partial x'^{i}\partial x'^{j}\partial x'^{k}}$ from (6.3.3) and (6.3.2),

$$\frac{\partial x^\alpha}{\partial x'^p}\left\{\frac{\partial}{\partial x'^j}\left(\Gamma'^p_{ik}\right)-\frac{\partial}{\partial x'^k}\left(\Gamma'^p_{ij}\right)\right\}+\Gamma'^p_{ik}\frac{\partial^2 x^\alpha}{\partial x'^j\,\partial x'^p}-\Gamma'^p_{ij}\frac{\partial^2 x^\alpha}{\partial x'^k\,\partial x'^p}$$

$$=\frac{\partial}{\partial x^b}\left(\Gamma^\alpha_{ac}\right)\frac{\partial x^a}{\partial x'^i}\frac{\partial x^b}{\partial x'^j}\frac{\partial x^c}{\partial x'^k}-\frac{\partial}{\partial x^c}\left(\Gamma^\alpha_{ab}\right)\frac{\partial x^a}{\partial x'^i}\frac{\partial x^b}{\partial x'^j}\frac{\partial x^c}{\partial x'^k}$$

$$+\Gamma^\alpha_{ac}\frac{\partial^2 x^a}{\partial x'^i\,\partial x'^j}\frac{\partial x^c}{\partial x'^k}-\Gamma^\alpha_{ab}\frac{\partial^2 x^a}{\partial x'^i\,\partial x'^k}\frac{\partial x^b}{\partial x'^j}+\Gamma^\alpha_{ab}\frac{\partial x^a}{\partial x'^i}\frac{\partial^2 x^b}{\partial x'^k\,\partial x'^j}-\Gamma^\alpha_{ab}\frac{\partial x^a}{\partial x'^i}\frac{\partial^2 x^b}{\partial x'^j\,\partial x'^k}$$

$$\therefore\ \frac{\partial x^\alpha}{\partial x'^p}\left[\frac{\partial}{\partial x'^j}\left(\Gamma'^p_{ik}\right)-\frac{\partial}{\partial x'^k}\left(\Gamma'^p_{ij}\right)\right]+\Gamma'^p_{ik}\left[\Gamma^m_{jp}\underset{(p\leftrightarrow m)}{\frac{\partial x^\alpha}{\partial x'^m}}-\Gamma^\alpha_{\lambda\mu}\frac{\partial x^\lambda}{\partial x'^j}\underset{(\lambda\to b)\ (\mu\to a)}{\frac{\partial x^\mu}{\partial x'^p}}\right]$$

$$-\Gamma'^p_{ij}\left[\Gamma'^m_{kp}\underset{(p\leftrightarrow m)}{\frac{\partial x^\alpha}{\partial x'^m}}-\Gamma^\alpha_{\lambda\mu}\frac{\partial x^\lambda}{\partial x'^k}\underset{(\lambda\to c)\ (\mu\to a)}{\frac{\partial x^\mu}{\partial x'^p}}\right]$$

$$=\left[\frac{\partial}{\partial x^b}\left(\Gamma^\alpha_{ac}\right)-\frac{\partial}{\partial x^c}\left(\Gamma^\alpha_{ab}\right)\right]\frac{\partial x^a}{\partial x'^i}\frac{\partial x^b}{\partial x'^j}\frac{\partial x^c}{\partial x'^k}+\Gamma^\alpha_{ac}\frac{\partial x^c}{\partial x'^k}\left[\Gamma'^m_{ij}\underset{(m\to p)}{\frac{\partial x^a}{\partial x'^m}}-\Gamma^a_{\lambda\mu}\frac{\partial x^\lambda}{\partial x'^i}\frac{\partial x^\mu}{\partial x'^j}\right]$$

$$-\Gamma^\alpha_{ab}\frac{\partial x^b}{\partial x'^j}\left[\Gamma'^m_{ik}\underset{(m\to p)}{\frac{\partial x^a}{\partial x'^m}}-\Gamma^a_{\lambda\mu}\frac{\partial x^\lambda}{\partial x'^i}\frac{\partial x^\mu}{\partial x'^k}\right]$$

$$\left[\frac{\partial}{\partial x'^j}\left(\Gamma'^p_{ik}\right)-\frac{\partial}{\partial x'^k}\left(\Gamma'^p_{ij}\right)+\Gamma'^m_{ik}\Gamma'^p_{jm}-\Gamma'^m_{ij}\Gamma'^p_{km}\right]\frac{\partial x^\alpha}{\partial x'^p}$$

$$=\left[\frac{\partial}{\partial x^b}\left(\Gamma^\alpha_{ac}\right)-\frac{\partial}{\partial x^c}\left(\Gamma^\alpha_{ab}\right)\right]\frac{\partial x^a}{\partial x'^i}\frac{\partial x^b}{\partial x'^j}\frac{\partial x^c}{\partial x'^k}-\Gamma^\alpha_{ac}\frac{\partial x^c}{\partial x'^k}\Gamma^a_{\lambda\mu}\underset{(\lambda\leftrightarrow a,\ \mu\to b)}{\frac{\partial x^\lambda}{\partial x'^i}\frac{\partial x^\mu}{\partial x'^j}}$$

$$+\Gamma^\alpha_{ab}\frac{\partial x^b}{\partial x'^j}\Gamma^a_{\lambda\mu}\underset{(\lambda\leftrightarrow a,\ \mu\to c)}{\frac{\partial x^\lambda}{\partial x'^i}\frac{\partial x^\mu}{\partial x'^k}}$$

$$\therefore R'^p_{ijk}\frac{\partial x^\alpha}{\partial x'^p}=\left[\frac{\partial}{\partial x^b}\left(\Gamma^\alpha_{ac}\right)-\frac{\partial}{\partial x^c}\left(\Gamma^\alpha_{ab}\right)-\Gamma^\alpha_{\lambda c}\Gamma^\lambda_{ab}+\Gamma^\alpha_{\lambda b}\Gamma^\lambda_{ac}\right]\times\frac{\partial x^a}{\partial x'^i}\frac{\partial x^b}{\partial x'^j}\frac{\partial x^c}{\partial x'^k}$$

$$R'^p_{ijk}\frac{\partial x^\alpha}{\partial x'^p}=R^\alpha_{abc}\frac{\partial x^\alpha}{\partial x'^i}\frac{\partial x^b}{\partial x'^j}\frac{\partial x^c}{\partial x'^k},$$

where $$R^\alpha_{abc}=\frac{\partial}{\partial x^b}\left(\Gamma^\alpha_{ac}\right)-\frac{\partial}{\partial x^c}\left(\Gamma^\alpha_{ab}\right)+\Gamma^\alpha_{\lambda b}\Gamma^\lambda_{ac}-\Gamma^\alpha_{\lambda c}\Gamma^\lambda_{ab}. \qquad (6.3.4)$$

Multiplying both sides by $\frac{\partial x'^h}{\partial x^\alpha}$, we get

$$R'^p_{ijk}\frac{\partial x^\alpha}{\partial x'^p}\frac{\partial x'^h}{\partial x^\alpha} = R^\alpha_{abc}\frac{\partial x^a}{\partial x'^i}\frac{\partial x^b}{\partial x'^j}\frac{\partial x^c}{\partial x'^k}\frac{\partial x'^h}{\partial x^\alpha}$$

or

$$R'^p_{ijk}\,\delta'^h_p = R^\alpha_{abc}\frac{\partial x^a}{\partial x'^i}\frac{\partial x^b}{\partial x'^j}\frac{\partial x^c}{\partial x'^k}\frac{\partial x'^h}{\partial x^\alpha}$$

$$\therefore R'^h_{ijk} = R^\alpha_{abc}\frac{\partial x^a}{\partial x'^i}\frac{\partial x^b}{\partial x'^j}\frac{\partial x^c}{\partial x'^k}\frac{\partial x'^h}{\partial x^\alpha},$$

which is the transformation law of the fourth-order mixed tensor.

R^α_{abc} in (6.3.4) is the similar expression for the Riemannian tensor or curvature tensor of the second kind.

6.4 Properties of the Curvature Tensor R^α_{ijk}

i. $R^\alpha_{ijk} = -R^\alpha_{ikj}$ (antisymmetric with respect to the second pair of indices).
ii. $R^\alpha_{ijk} + R^\alpha_{jki} + R^\alpha_{kij} = 0$.
iii. Contraction in two different ways: (a) $R^i_{ijk} = 0$ and (b) $R^\alpha_{ij\alpha}$, a tensor called Ricci tensor.

Proof of (i)

By definition,

$$R^\alpha_{ijk} = \frac{\partial}{\partial x^j}\left(\Gamma^\alpha_{ik}\right) - \frac{\partial}{\partial x^k}\left(\Gamma^\alpha_{ij}\right) + \Gamma^\alpha_{bj}\Gamma^b_{ik} - \Gamma^\alpha_{bk}\Gamma^b_{ij}$$

and

$$R^\alpha_{ikj} = \frac{\partial}{\partial x^k}\left(\Gamma^\alpha_{ij}\right) - \frac{\partial}{\partial x^j}\left(\Gamma^\alpha_{ik}\right) + \Gamma^\alpha_{bk}\Gamma^b_{ij} - \Gamma^\alpha_{bj}\Gamma^b_{ik}.$$

On comparison, it is clear that $R^\alpha_{ijk} = -R^\alpha_{ikj}$, i.e., antisymmetric with respect to j and k.

Hence, proved.

Proof of (ii)

By definition, it can be written as

$$R^{\alpha}_{ijk}+R^{\alpha}_{jki}+R^{\alpha}_{kij}=\frac{\partial}{\partial x^j}\left(\Gamma^{\alpha}_{ik}\right)-\frac{\partial}{\partial x^k}\left(\Gamma^{\alpha}_{ij}\right)+\Gamma^{\alpha}_{bj}\Gamma^{b}_{ik}-\Gamma^{\alpha}_{bk}\Gamma^{b}_{ij}$$

$$+\frac{\partial}{\partial x^k}\left(\Gamma^{\alpha}_{ji}\right)-\frac{\partial}{\partial x^i}\left(\Gamma^{\alpha}_{jk}\right)+\Gamma^{\alpha}_{bk}\Gamma^{b}_{ji}-\Gamma^{\alpha}_{bi}\Gamma^{b}_{jk}+\frac{\partial}{\partial x^i}\left(\Gamma^{\alpha}_{kj}\right)$$

$$-\frac{\partial}{\partial x^j}\left(\Gamma^{\alpha}_{ki}\right)+\Gamma^{\alpha}_{bi}\Gamma^{b}_{kj}-\Gamma^{\alpha}_{bj}\Gamma^{b}_{ki}$$

$$=0 \text{ since } \Gamma^{b}_{ij}=\Gamma^{b}_{ji}.$$

It is called cyclic property.

Hence, proved.

Proof of (iii)

a. Considering a contraction with the first index **in** R^{α}_{ijk}, we get (i.e., $\alpha=i$)

$$R^{i}_{ijk}=\frac{\partial}{\partial x^j}\left(\Gamma^{i}_{ik}\right)-\frac{\partial}{\partial x^k}\left(\Gamma^{i}_{ij}\right)+\Gamma^{i}_{bj}\Gamma^{b}_{ik}-\Gamma^{i}_{bk}\Gamma^{b}_{ij}\ (b\leftrightarrow i)$$

$$=\frac{\partial}{\partial x^j}\left[\frac{\partial}{\partial x^k}\left(\log\sqrt{g}\right)\right]-\frac{\partial}{\partial x^k}\left[\frac{\partial}{\partial x^j}\left(\log\sqrt{g}\right)\right],\ g>0$$

$$=\frac{\partial^2}{\partial x^j\,\partial x^k}\left(\log\sqrt{g}\right)-\frac{\partial^2}{\partial x^k\,\partial x^j}\left(\log\sqrt{g}\right)=0$$

$$\because\ \Gamma^{i}_{ij}=\frac{\partial}{\partial x^j}\left(\log\sqrt{\pm g}\right).$$

b. Considering a contraction with the third index in R^{α}_{ijk}, we get (i.e., $\alpha=k$)

$$R^{\alpha}_{ij\alpha}=\frac{\partial}{\partial x^j}\left(\Gamma^{\alpha}_{i\alpha}\right)-\frac{\partial}{\partial x^\alpha}\left(\Gamma^{\alpha}_{ij}\right)+\Gamma^{\alpha}_{bj}\Gamma^{b}_{i\alpha}-\Gamma^{\alpha}_{b\alpha}\Gamma^{b}_{ij}$$

$$=\frac{\partial}{\partial x^j}\left[\frac{\partial}{\partial x^i}\left(\log\sqrt{g}\right)\right]-\frac{\partial}{\partial x^\alpha}\left(\Gamma^{\alpha}_{ij}\right)+\Gamma^{\alpha}_{bj}\Gamma^{b}_{i\alpha}-\Gamma^{b}_{ij}\frac{\partial}{\partial x^b}\left(\log\sqrt{g}\right),\quad \text{when } g>0$$

$$\therefore R_{ij}=\frac{\partial^2}{\partial x^i\,\partial x^j}\left(\log\sqrt{g}\right)-\frac{\partial}{\partial x^\alpha}\left(\Gamma^{\alpha}_{ij}\right)+\Gamma^{\alpha}_{bj}\Gamma^{b}_{i\alpha}-\Gamma^{b}_{ij}\frac{\partial}{\partial x^b}\left(\log\sqrt{g}\right).$$

This contracted tensor is called the **Ricci tensor**. In future, it will have a **great deal of application** in the field equation of general theory of relativity.

It can also be easily seen that $R_{ij}=R_{ji}$.

6.5 Covariant Curvature Tensor

The fourth-order covariant curvature tensor (by virtue of definition R^{α}_{ijk}) is defined as

$$R_{hijk} = g_{ha} R^{a}_{ijk}. \tag{6.5.1}$$

The symbols R_{hijk} are known as the Riemannian symbols of the **first kind**.

Now,

$$R_{hijk} = g_{ha}\left[\frac{\partial}{\partial x^j}\left(\Gamma^{a}_{ik}\right) - \frac{\partial}{\partial x^k}\left(\Gamma^{a}_{ij}\right) + \Gamma^{a}_{bj}\Gamma^{b}_{ik} - \Gamma^{a}_{bk}\Gamma^{b}_{ij}\right]$$

$$= \frac{\partial}{\partial x^j}\left(g_{ha}\Gamma^{a}_{ik}\right) - \frac{\partial}{\partial x^k}\left(g_{ha}\Gamma^{a}_{ij}\right) - \frac{\partial g_{ha}}{\partial x^j}\Gamma^{a}_{ik} + \frac{\partial g_{ha}}{\partial x^k}\Gamma^{a}_{ij} + g_{ha}\Gamma^{a}_{bj}\Gamma^{b}_{ik} - g_{ha}\Gamma^{a}_{bk}\Gamma^{b}_{ij}$$

$$= \frac{\partial}{\partial x^j}\left(\Gamma_{ik,h}\right) - \frac{\partial}{\partial x^k}\left(\Gamma_{ij,h}\right) - \left(\Gamma_{hj,a} + \Gamma_{aj,h}\right)\Gamma^{a}_{ik} + \left(\Gamma_{hk,a} + \Gamma_{ak,h}\right)\Gamma^{a}_{ij} + \Gamma_{bj,h}\Gamma^{b}_{ik} - \Gamma_{bk,h}\Gamma^{b}_{ij}$$

$$R_{hijk} = \frac{1}{2}\frac{\partial}{\partial x^j}\left[\frac{\partial g_{ih}}{\partial x^k} + \frac{\partial g_{kh}}{\partial x^i} - \frac{\partial g_{ik}}{\partial x^h}\right] - \frac{1}{2}\frac{\partial}{\partial x^k}\left[\frac{\partial g_{ih}}{\partial x^j} + \frac{\partial g_{jh}}{\partial x^i} - \frac{\partial g_{ij}}{\partial x^h}\right]$$

$$- g_{\alpha a}\Gamma^{\alpha}_{hj}\Gamma^{a}_{ik} - g_{\alpha h}\Gamma^{\alpha}_{aj}\Gamma^{a}_{ik} + g_{\alpha a}\Gamma^{\alpha}_{hk}\Gamma^{a}_{ij} + g_{\alpha h}\Gamma^{\alpha}_{ak}\Gamma^{a}_{ij}$$

$$+ g_{\alpha h}\underset{(b\to a)}{\Gamma^{\alpha}_{bj}}\Gamma^{b}_{ik} - g_{\alpha h}\underset{(b\to a)}{\Gamma^{\alpha}_{bk}}\Gamma^{b}_{ij}$$

$$= \frac{1}{2}\left[\frac{\partial^2 g_{ih}}{\partial x^j \partial x^k} + \frac{\partial^2 g_{kh}}{\partial x^j \partial x^i} - \frac{\partial^2 g_{ik}}{\partial x^j \partial x^h} - \frac{\partial^2 g_{ih}}{\partial x^k \partial x^j} - \frac{\partial^2 g_{jh}}{\partial x^k \partial x^i} + \frac{\partial^2 g_{ij}}{\partial x^k \partial x^h}\right]$$

$$+ g_{\alpha a}\Gamma^{\alpha}_{hk}\Gamma^{a}_{ij} - g_{\alpha a}\Gamma^{\alpha}_{hj}\Gamma^{a}_{ik}$$

$$= \frac{1}{2}\left[\frac{\partial^2 g_{ij}}{\partial x^h \partial x^k} + \frac{\partial^2 g_{kh}}{\partial x^i \partial x^j} - \frac{\partial^2 g_{ik}}{\partial x^j \partial x^h} - \frac{\partial^2 g_{jh}}{\partial x^i \partial x^k}\right]$$

$$+ g_{\alpha a}\Gamma^{\alpha}_{hk}\Gamma^{a}_{ij} - g_{\alpha a}\Gamma^{\alpha}_{hj}\Gamma^{a}_{ik}, \tag{6.5.2}$$

which is the expression of the **curvature tensor** of the first kind.

6.6 Properties of the Curvature Tensor R_{hijk} of the First Kind

i. $R_{hijk} = -R_{ihjk}$.
ii. $R_{hijk} = -R_{hikj}$ (antisymmetric with respect to the second pair of indices).
iii. $R_{hijk} = R_{jkhi}$ (symmetric with respect to the first and second pairs of indices).
iv. $R_{iijk} = R_{hikk} = 0$.
v. $R_{hijk} + R_{hjki} + R_{hkij} = 0$.

By definition (6.5.2), we get

$$R_{hijk} = \frac{1}{2}\left[\frac{\partial^2 g_{ij}}{\partial x^h \partial x^k} + \frac{\partial^2 g_{hk}}{\partial x^i \partial x^j} - \frac{\partial^2 g_{ik}}{\partial x^h \partial x^j} - \frac{\partial^2 g_{hj}}{\partial x^i \partial x^k}\right] + g_{\alpha a}\Gamma^{\alpha}_{hk}\Gamma^{a}_{ij} - g_{\alpha a}\Gamma^{\alpha}_{ik}\Gamma^{a}_{hj}.$$

Interchanging h and i,

$$R_{ihjk} = \frac{1}{2}\left[\frac{\partial^2 g_{hj}}{\partial x^i \partial x^k} + \frac{\partial^2 g_{ik}}{\partial x^h \partial x^j} - \frac{\partial^2 g_{hk}}{\partial x^i \partial x^j} - \frac{\partial^2 g_{ij}}{\partial x^h \partial x^k}\right]$$

$$+ g_{\alpha a}\Gamma^{\alpha}_{ik}\Gamma^{a}_{hj} - g_{\alpha a}\Gamma^{\alpha}_{ij}\Gamma^{a}_{hk} \qquad (\alpha \leftrightarrow a).$$

Comparison gives the following result:

i. $R_{hijk} = -R_{ihjk}$, i.e., antisymmetric with respect to first pair of indices.
ii. Similarly, it can be proved $R_{hijk} = -R_{hikj}$, i.e., antisymmetric with respect to second pair of indices.
iii. $R_{hijk} = R_{jkhi}$, i.e., symmetric with respect to the first and second pair of indices.

Proof

From definition (6.5.2), it follows

$$R_{hijk} = \frac{1}{2}\left[\frac{\partial^2 g_{ij}}{\partial x^h \partial x^k} + \frac{\partial^2 g_{hk}}{\partial x^i \partial x^j} - \frac{\partial^2 g_{ik}}{\partial x^h \partial x^j} - \frac{\partial^2 g_{hj}}{\partial x^i \partial x^k}\right] + g_{\alpha a}\Gamma^{\alpha}_{hk}\Gamma^{a}_{ij} - g_{\alpha a}\Gamma^{\alpha}_{hj}\Gamma^{a}_{ik}$$

and

$$R_{jkhi} = \frac{1}{2}\left(\frac{\partial^2 g_{kh}}{\partial x^j \partial x^i} + \frac{\partial^2 g_{ji}}{\partial x^k \partial x^h} - \frac{\partial^2 g_{ki}}{\partial x^j \partial x^h} - \frac{\partial^2 g_{jh}}{\partial x^k \partial x^i}\right) + g_{\alpha a}\Gamma^{\alpha}_{ij}\Gamma^{a}_{kh}$$

$$- g_{\alpha a}\Gamma^{\alpha}_{ki}\Gamma^{a}_{jh} \qquad (\alpha \leftrightarrow a).$$

Clearly, $R_{hijk} = R_{jkhi} \therefore \Gamma_{ij}^{\alpha} = \Gamma_{ji}^{\alpha}$ and $\frac{\partial^2 g_{ik}}{\partial x^h \partial x^j} = \frac{\partial^2 g_{ki}}{\partial x^j \partial x^h}$, i.e., symmetric with respect to the two pairs of indices.

Hence, proved.

iv. $R_{iijk} = R_{hikk} = 0.$

Replacing k by i and j by k, it can easily be proved from definition.

v. $R_{hijk} + R_{hjki} + R_{hkij} = 0$

Using the definition of R^{α}_{ijk}, it is proved in Section 6.4 (ii) that $R^{\alpha}_{ijk} + R^{\alpha}_{jki} + R^{\alpha}_{kij} = 0$, the cyclic property.

Multiplying it by $g_{h\alpha}$ and summing over α,

$$g_{h\alpha} R^{\alpha}_{ijk} + g_{h\alpha} R^{\alpha}_{jki} + g_{h\alpha} R^{\alpha}_{kij} = 0$$

$$\therefore R_{hijk} + R_{hjki} + R_{hkij} = 0.$$

Hence, proved.

6.7 Bianchi Identity

Let P_o be the pole of a geodesic coordinates x^i for which $\Gamma_{jk}^{i} = 0 = \Gamma_{ij,k}$ and so $A_{i,j} = \frac{\partial A_i}{\partial x_j}$, A_i is any tensor (vector).

Now from definition of curvature tensor of the second kind, we have

$$R^a_{ijk} = \frac{\partial}{\partial x^j}\left(\Gamma_{ik}^{a}\right) - \frac{\partial}{\partial x^k}\left(\Gamma_{ij}^{a}\right) + \Gamma_{\alpha j}^{a} \Gamma_{ik}^{\alpha} - \Gamma_{\alpha k}^{a} \Gamma_{ij}^{\alpha}. \tag{6.7.1}$$

Considering x^l-covariant differentiation of (6.7.1) with respect to g_{ij}, we get

$$R^a_{ijk,l} = \frac{\partial^2}{\partial x^l \partial x^j}\left(\Gamma_{ik}^{a}\right) - \frac{\partial^2}{\partial x^l \partial x^k}\left(\Gamma_{ij}^{a}\right) \quad \text{at pole } P_o$$

Similarly,

$$R^a_{ikl,j} = \frac{\partial^2}{\partial x^j \partial x^k}\left(\Gamma_{il}^{a}\right) - \frac{\partial^2}{\partial x^j \partial x^l}\left(\Gamma_{ik}^{a}\right) \quad \text{at pole } P_o$$

and

$$R^a_{ilj,k} = \frac{\partial^2}{\partial x^k \partial x^l}\left(\Gamma_{ij}^{a}\right) - \frac{\partial^2}{\partial x^k \partial x^j}\left(\Gamma_{il}^{a}\right) \quad \text{at pole } P_o.$$

Addition of all these three relations gives

$$R^{a}_{ijk,l} + R^{a}_{ikl,j} + R^{a}_{ilj,k} = 0 \quad \text{at pole } P_o. \tag{6.7.2}$$

Each of the terms of the equation is a tensor, so it holds for all coordinate systems and at all points. Hence, it is an identity instead of an equation. It is known as **Bianchi identity.**

The inner product of (6.7.2) with g_{ha} (summing over a) gives

$$\left(g_{ha}R^{a}_{ijk}\right)_{,l} + \left(g_{ha}R^{a}_{ikl}\right)_{,j} + \left(g_{ha}R^{a}_{ilj}\right)_{,k} = 0 \because g_{ha,k} = 0, \text{ etc.}$$

$$R_{hijk,l} + R_{hikl,j} + R_{hilj,k} = 0.$$

It is an alternative form of the **Bianchi identity.**

6.8 Einstein Tensor Is Divergence Free

The Bianchi identity $R^{a}_{ijk,l} + R^{a}_{ikl,j} + R^{a}_{ilj,k} = 0$ can be written as

$$R^{a}_{ijk,l} - R^{a}_{ilk,j} + R^{a}_{ilj,k} = 0\left(R^{a}_{ikl,j} = -R^{a}_{ilk,j}\right).$$

Considering a contraction with respect to a and k,

$$R^{a}_{ija,l} - R^{a}_{ila,j} + R^{a}_{ilj,a} = 0$$

$$\left(R^{a}_{ija} = R_{ij}, R_{ii} \text{ is Ricci tensor}\right).$$

An inner multiplication of it with g^{il} gives

$$\left(g^{il}R_{ij}\right)_{,l} - \left(g^{il}R_{il}\right)_{,j} + \left(g^{il}R^{a}_{ilj}\right)_{,a} = 0$$

$$\because g^{il} \text{ is covariant constant}$$

$$R^{l}_{j,l} - R_{,j} + R^{a}_{j,a} = 0.$$

It can be written as $2R^{i}_{j,i} - R_{,j} = 0$, changing $l \to i$ and $a \to i$.

$$\therefore \left(R^{i}_{j} - \frac{1}{2}\delta^{i}_{j}R\right)_{,i} = 0.$$

Hence, $G^i_{j,i} = 0$, where $G^i_j = R^i_j - \frac{1}{2}\delta^i_j R$ is the **Einstein tensor**.

∴ The Einstein tensor is divergence free (∵ contraction is with respect to covariant derivative index which is required for definition of divergence).

6.9 Isometric Surfaces

The intrinsic geometry of a surface is based on the corresponding fundamental quadratic form or metric $ds^2 = a_{\alpha\beta} du^\alpha u^\beta$ with surface coordinates $u^\alpha = u^\alpha(u^1, u^2)$. The intrinsic properties such as the lengths of curves and the angle between two intersecting curves primarily depend on the metric tensor of the surface and its derivatives. If there exists a coordinate system which characterizes the linear element of two surfaces S_1 and S_2 by the same metric $a_{\alpha\beta}$, then they are called isometric. The corresponding parameters for the transformation is known as isometry. From the Euclidean plane, surfaces of cylinder and cone can be constructed by means of rolling without changing arc length, areas, and measurement of angles. Hence, they are isometric with the Euclidean plane.

6.10 Three-Dimensional Orthogonal Cartesian Coordinate Metric and Two-Dimensional Curvilinear Coordinate Surface Metric Imbedded in It

In order to enter into the threshold of geometry of surfaces in a surrounding space, we need to consider two distinct coordinate systems. Let $u^\alpha(u^1, u^2)$ be the two curvilinear coordinates of the surface S imbedded in a three orthogonal Cartesian coordinate systems in E_3. But the intrinsic property of geometry of a space is characterized by the metric or quadratic differential form.

Let $x^i = x^i(y^1, y^2, y^3)$, $i = 1, 2, 3$, be the orthogonal Cartesian coordinates covering the space E_3 and $x^i = x^i(u^1, u^2)$ be the Gaussian surface coordinates of S imbedded in E_3. The line element in E_3 is given by $ds^2 = g_{ij} dx^i dx^j$, where

$$dx^i = \frac{\partial x^i}{\partial u^\alpha} du^\alpha \text{ and } g_{ij} = \frac{\partial y^k}{\partial x^i}\frac{\partial y^k}{\partial x^j}$$

$$= g_{ij} \frac{\partial x^i}{\partial u^\alpha} du^\alpha \frac{\partial x^j}{\partial u^\beta} du^\beta$$

$$ds^2 = g_{ij}\frac{\partial x^i}{\partial u^\alpha}\frac{\partial x^j}{\partial u^\beta}du^\alpha du^\beta = a_{\alpha\beta}du^\alpha du^\beta,$$

where (6.10.1)

$$a_{\alpha\beta} = g_{ij}\frac{\partial x^i}{\partial u^\alpha}\frac{\partial x^j}{\partial u^\beta},$$

so that $a = gJ^2, |a_{\alpha\beta}| = a, \quad |g_{ij}| = g.$

Looking at $dx^i = \frac{\partial x^i}{\partial u^\alpha}du^\alpha$, it can be concluded that dx^i is a space vector and is surface invariant, and du^α is a surface vector and is space invariant.

6.11 Gaussian Curvature of the Surface S immersed in E_3

If the functions $a_{\alpha\beta}$ and $b_{\alpha\beta}\left(= g_{ij}\right)$ in (6.10.1) related to some surface, then x^i is to satisfy the condition of integrability:

$$\frac{\partial^2 x^i_\alpha}{\partial u^\beta\,\partial u^\gamma} = \frac{\partial^2 x^i_\alpha}{\partial u^\gamma\,\partial u^\beta}, \tag{6.11.1}$$

where $x^i_\alpha = \frac{\partial x^i}{\partial u^\alpha}$ is continuously differentiable of degree 2.

Now denoting $\frac{\partial x^i}{\partial u^\alpha}$ by t^i_α which is tangent to the surface and considering the covariant (surface) derivative of it, we can get

$$t^i_{\alpha,\beta} = \frac{\partial t^i_\alpha}{\partial u^\beta} + \Gamma^i_{jk}\,t^j_\alpha\,t^k_\beta - \Gamma^\gamma_{\alpha\beta}\,t^i_\gamma = \frac{\partial^2 x^i}{\partial u^\alpha\,\partial u^\beta} + \Gamma^i_{jk}\,t^j_\alpha\,t^k_\beta - \Gamma^\gamma_{\alpha\beta}\,t^i_\gamma.$$

Since t^i_α is tangent to the surface, its partial derivative $t^i_{\alpha,\beta}$ is normal to the surface; otherwise, it is proportional to the normal n^i of the surface. Therefore, $t^i_{\alpha,\beta} = b_{\alpha\beta}\,n^i$

or

$$b_{\alpha\beta} = t^i_{\alpha,\beta}\,n^i = \left[\frac{\partial^2 x^i}{\partial u^\alpha\,\partial u^\beta} + \Gamma^i_{jk}\,t^j_\alpha\,t^k_\beta - \Gamma^\gamma_{\alpha\beta}\,t^i_\gamma\right]n_i. \tag{6.11.2}$$

But for the use of Cartesian coordinates and geodesic surface coordinates, the Christoffel symbols can be made to zero at a particular point. Of course, the derivatives of the Christoffel symbols in space $\left(\Gamma^i_{jk}\right)$ will vanish but not the symbols with surface $\left(\Gamma^\gamma_{\alpha\beta}\right)$.

Hence, $t^i_{\alpha,\beta\gamma} = \dfrac{\partial^3 x^i}{\partial u^\alpha \partial u^\beta \partial u^\gamma} - \dfrac{\partial}{\partial u^\gamma}\left[\Gamma^{\epsilon}_{\alpha\beta}\right] t^i_{\epsilon}.$

Similarly, $t^i_{\alpha,\gamma\beta} = \dfrac{\partial^3 x^i}{\partial u^\alpha \partial u^\gamma \partial u^\beta} - \dfrac{\partial}{\partial u^\beta}\left[\Gamma^{\epsilon}_{\gamma\alpha}\right] t^i_{\epsilon}.$

$$\therefore t^i_{\alpha,\beta\gamma} - t^i_{\alpha,\gamma\beta} = R^{\epsilon}_{\alpha\beta\gamma}\, t^i_{\epsilon}. \tag{6.11.3}$$

Using (6.11.1) and (6.11.2) and making use of definition for $R^{\epsilon}_{\alpha\beta\gamma}$

From $t^i_{\alpha,\,\beta} = b_{\alpha\beta} n^i$, we can get

$$t^i_{\alpha,\,\beta\gamma} = b_{\alpha\beta,\,\gamma} n^i + b_{\alpha\beta} n^i_{\,,\,\gamma}$$
$$= b_{\alpha\beta,\,\gamma} n^i - b_{\alpha\beta} b_{\gamma\delta} a^{\delta\epsilon} t^i_{\,\varepsilon}.$$

($\because n^i_{\,,\,\alpha} = -\, b_{\alpha\beta} a^{\beta\gamma} t^i_{\,\gamma}$ from Weingarten's formula due to the linear combination of n^i_α and t^i_β, namely, $n^i_{,\,\alpha} = C^{\beta}_{\alpha} t^i_{\beta}$)

$$\therefore t^i_{\alpha,\,\beta\gamma} - t^i_{\alpha,\,\gamma\beta} = \left(b_{\alpha\beta,\,\gamma} - b_{\alpha\gamma,\,\beta}\right) n^i + \left(b_{\alpha\gamma} b_{\beta\delta} - b_{\alpha\beta} b_{\gamma\delta}\right) a^{\delta\epsilon} t^i_{\varepsilon}.$$

$$\therefore R^{\varepsilon}_{\alpha\beta\gamma} t^i_{\varepsilon} = \left(b_{\alpha\beta,\,\gamma} - b_{\alpha\gamma,\,\beta}\right) n^i + \left(b_{\alpha\gamma} b_{\beta\delta} - b_{\alpha\beta} b_{\gamma\delta}\right) a^{\delta\epsilon} t^i_{\varepsilon}, \quad \text{using (6.11.3)}$$

$$\therefore b_{\alpha\beta,\,\gamma} - b_{\alpha\gamma,\,\beta} = 0, \tag{6.11.4}$$

multiplying by n^i and using $n^i t^i_{\varepsilon} = 0$.

This is known as **Codazzi equation** of surface. Putting this value, the above equation can be reduced to

$$R^{\varepsilon}_{\alpha\beta\gamma} t^i_{\varepsilon} = \left(b_{\alpha\gamma} b_{\beta\delta} - b_{\alpha\beta} b_{\gamma\delta}\right) a^{\delta\epsilon} t^i_{\varepsilon}.$$

Therefore, for arbitrary t^i_{ε}, it can be transformed to

$$R_{\lambda\alpha\beta\gamma} = b_{\alpha\gamma} b_{\beta\lambda} - b_{\alpha\beta} b_{\gamma\lambda}. \tag{6.11.5}$$

$$\because a_{\lambda\epsilon} a^{\delta\epsilon} = \delta^{\delta}_{\lambda}.$$

This is known as **Gauss equation** of surface.

But for Riemannian curvature tensor $R_{\lambda\alpha\beta\gamma}$, $R_{\alpha\alpha\beta\gamma} = R_{\lambda\alpha\beta\beta} = 0$, the only surviving component is $R_{1212} = -R_{2112} = -R_{1221} = R_{2121}$.

Since the surface $x^i = x^i\left(u^1,\, u^2\right)$ in two-dimensional curvilinear coordinates is immersed in E_3 with the metric $a_{\alpha\beta}$ or $\left|a_{\alpha\beta}\right| = a$, the quantity defined by $\kappa = \dfrac{R_{1212}}{a}$ is called the total curvature or the Gaussian curvature.

But $R_{1212} = b_{11} b_{22} - b_{12} b_{21} = b_{11} b_{22} - b^2_{12} = b$, from (6.11.5) so that

$$\kappa = \frac{b_{11}b_{22} - b_{12}^2}{a}\left(= \frac{b}{a}\right)$$

$$= \frac{b_{11}b_{22} - b_{12}^2}{a_{11}a_{22} - a_{12}^2}$$

is the **Gaussian curvature** of two-dimensional surface.

Example 1

Prove that the differential equation $A_{i,j} = 0$ is integrable only when the Riemann Christoffel tensor vanishes.

It needs to show that $R^a_{ijk} = 0$, if the differential equation $A_{i,j} = 0$ is integrable.

Now,

$$A_{i,j} = 0 \text{ gives } \quad \frac{\partial A_i}{\partial x^j} - A_a \Gamma_{ij}^a = 0. \tag{i}$$

$$\frac{\partial A_i}{\partial x^j} dx^j = A_a \Gamma_{ij}^a dx^j \quad \therefore A_i = \int A_a \Gamma_{ij}^a dx^j. \tag{ii}$$

This shows that the right-hand side of (ii) must be integrable, and hence, it should be a perfect differential of some function, say B_i, so that

$$A_a \Gamma_{ij}^a dx^j = dB_i$$

$$\therefore \frac{\partial B_i}{\partial x^j} dx^j = A_a \Gamma_{ij}^a dx^j$$

$$\left(\frac{\partial B_i}{\partial x^j} - A_a \Gamma_{ij}^a\right) dx^j = 0.$$

It gives $\frac{\partial B_i}{\partial x^j} = A_a \Gamma_{ij}^a \because dx^j$ is arbitrary.

Differentiating partially with respect to x^k, we get

$$\frac{\partial^2 B}{\partial x^j \partial x^k} = \frac{\partial A_a}{\partial x^k} \Gamma_{ij}^a + A_a \frac{\partial}{\partial x^k}\left(\Gamma_{ij}^a\right). \tag{iii}$$

Interchanging j and k, it can be written as

$$\frac{\partial^2 B}{\partial x^k \partial x^j} = \frac{\partial A_a}{\partial x^j} \Gamma_{ik}^a + A_a \frac{\partial}{\partial x^j}\left(\Gamma_{ik}^a\right). \tag{iv}$$

From (iii) and (iv),

$$\frac{\partial A_a}{\partial x^j} \Gamma_{ik}^a + A_a\left(\frac{\partial}{\partial x^j} \Gamma_{ik}^a\right) - \frac{\partial A_a}{\partial x^k} \Gamma_{ij}^a - A_a\left(\frac{\partial}{\partial x^k} \Gamma_{ij}^a\right) = 0$$

$$\underset{(a \leftrightarrow b)}{A_b \Gamma_{aj}^b \Gamma_{ik}^a} + A_a \frac{\partial}{\partial x^j}\left(\Gamma_{ik}^a\right) - \underset{(a \leftrightarrow b)}{A_b \Gamma_{ak}^b \Gamma_{ij}^a} - A_a \frac{\partial}{\partial x^k}\left(\Gamma_{ij}^a\right) = 0.$$

(making use of (i))

$$\therefore A_a\left[\frac{\partial}{\partial x^j}\left(\Gamma^a_{ik}\right)-\frac{\partial}{\partial x^k}\left(\Gamma^a_{ij}\right)+\Gamma^a_{bj}\Gamma^b_{ik}-\Gamma^a_{bk}\Gamma^b_{ij}\right]=0.$$

$A_aR^a_{ijk}=0 \therefore R^a_{ijk}=0 \because A_a$ is arbitrary in the inner product.

Hence, proved.

Example 2

Show that the number of independent components of the Riemannian curvature tensor of the first kind R_{hijk} in n-dimensional space V_n is $\frac{1}{2}n^2\left(n^2-1\right)$.

In general, the number of independent components of the fourth-order tensor R_{hijk} in a Riemannian space V_n is n^4. But due to the following properties, the number of independent components will be reduced from n^4.

i $R_{hijk}=-R_{ihjk}$ (antisymmetric property with respect to h and i).
ii $R_{hijk}=-R_{hikj}$ (antisymmetric property with respect to j and k).
iii $R_{hijk}=R_{jkhi}$ (symmetric with respect to the first and second pairs).
iv $R_{hijk}+R_{hjki}+R_{hkij}=0$ (cyclic property).

Case I: When there is only one distinct index of the type R_{hhhh}

$$R_{hhhh}=-R_{hhhh}\left(\text{due to }(i)\right)\quad \therefore R_{hhhh}=0.$$

Hence, there is no component of R_{hijk} with one distinct index.

Case II: When there are two distinct indices of the type $R_{hihi}\left(h\neq i\right)$

The two distinct indices from n values can be selected in $n(n-1)$ ways which correspond to $n(n-1)$ numbers of independent components of R_{hijk}.

But $R_{hihi}=-R_{ihhi}=R_{ihih}$, due to (i) and (ii).

After interchanging h and i in the first, the last one can be recovered.

$\therefore$ The number $n(n-1)$ is reduced only by $\frac{1}{2}$ so that it becomes $\frac{1}{2}n(n-1)$.

Also, $R_{hihi}=R_{hihi}$ (due to (iii)), so there is no reduction due to this property.

Interestingly, the cyclic property

$$R_{hihi}+R_{hhii}+R_{hiih}=R_{hihi}-R_{hihi}=0$$

is identically satisfied due to (ii).

Hence, there is no reduction due to the cyclic property.

$\therefore$ The number of independent components in this case is $\frac{1}{2}n(n-1)$.

Case III: When there are three distinct indices of the type R_{hihj}

$\therefore$ The values of h, i, j can be selected from n values in $n(n-1)(n-2)$ ways, and hence, the number of independent components of R_{hijk} in this case is $n(n-1)(n-2)$. But this number will be reduced for the properties (i)–(iv).

Now, $R_{hihj} = R_{hjhi}$ (due to (iii)) which can be recovered by merely interchanging i and j
$= -R_{jhhi}$ (due to (i))
$= -R_{hijh}$ (due to (iii)) which is nothing but (ii).

Hence, due to symmetric properties, the number of independent components is finally reduced by $\frac{1}{2}$ only to yield $\frac{1}{2}\, n(n-1)(n-2)$.

The cyclic property $R_{hihj} + R_{hhji} + R_{hjih} = R_{hihj} + 0 + R_{ihhj} = R_{hihj} - R_{hihj} = 0$ is identically satisfied, and hence, there is no reduction of the above number.

Case IV: When all the four indices are distinct of the type R_{hijk}

Clearly, the values of h, i, j, k can be selected in $n(n-1)(n-2)(n-3)$ ways, and hence, the number of independent components in this case is $n(n-1)(n-2)(n-3)$.

But due to the **three** symmetric properties, (i)–(iii), it is reduced to $\frac{1}{2}\times\frac{1}{2}\times\frac{1}{2}n(n-1)(n-2)(n-3)$.

But from (iv), we get $R_{hijk} + R_{hjki} = -R_{hkij}$.

This shows that, knowing two components, the third can be readily determined. Hence, due to (iv), the number of independent components is reduced by $\frac{2}{3}$.

$\therefore$ In this case, the number of independent components of R_{hijk} is $\frac{1}{8}\times\frac{2}{3}n(n-1)(n-2)(n-3)$.

Hence, the total number of independent components of the Riemannian curvature tensor R_{hijk} is

$$0+\frac{1}{2}n(n-1)+\frac{1}{2}n(n-1)(n-2)+\frac{1}{12}n(n-1)(n-2)(n-3)$$

$$=\frac{1}{12}n(n-1)\left[6+6(n-2)+(n-2)(n-3)\right]$$

$$=\frac{1}{12}n(n-1)\left[6n-6+n^2-5n+6\right]$$

$$=\frac{1}{12}n(n-1)n(n+1)=\frac{1}{12}n^2(n^2-1).$$

Hence, proved.

Exercises

1. Show that the divergence of the tensor $R^i_j - \frac{1}{2}\delta^i_j R$ is identically zero.
2. If the metric of V_2 formed by the surface of a sphere of radius r is $ds^2 = r^2\ (d\theta^2 + \sin 2\theta\ d\phi^2)$ in spherical polar coordinates, show that $R_{1212} = r^2 \sin^2\theta$.

3. Derive the expression of the curvature tensor of the second kind R^{a}_{ijk}.
4. Derive the expression of the curvature tensor of the first kind R_{hijk}.
5. Discuss the properties of R_{hijk}.
6. For a V_3 referred to a triply orthogonal coordinate system, prove that $R_{ij} = \dfrac{1}{g_{kk}} R_{ikkj}$ (where $i \neq j \neq k$)

   and $R_{hh} = \dfrac{1}{g_{ii}} R_{hiih} + \dfrac{1}{g_{jj}} R_{hjjh}$.
7. Calculate the Ricci tensor R_{ij} for the metric on the sphere $ds^2 = a^2 (d\theta^2 + \sin^2\theta \, d\phi^2)$, where $i, j = 1, 2$, $x^1 = \theta$ and $x^2 = \phi$, and a is constant.
   [Hint: Find Γ^{α}_{ij} and use $\Gamma^{i}_{ij} = \dfrac{\partial}{\partial x^j}(\log \sqrt{g})$.]
8. Show that on a two-dimensional surface, the curvature tensor is completely defined by a single component, say R_{1212}.
9. For a surface with the metric $ds^2 = (du)^2 + \lambda^2 (dv)^2$, show that the Gaussian curvature is $-\dfrac{1}{\lambda}\dfrac{\partial^2 \lambda}{\partial u^2}$.
10. Show that the surface with the metric $ds^2 = \left(u^2\right)^2 \left(du^1\right)^2 + \left(u^1\right)^2 \left(du^2\right)^2$ is isometric or developable.
    [Hint: Show $\kappa = 0$.]
11. Find the conditions that the surfaces $S_1 : y^1 = v^1 \cos v^2$, $y^2 = v^1 \sin v^2, y^3 = a \cosh^{-1}\left(\dfrac{v^1}{a}\right)$; $S_2 : y^1 = u^1 \cos u^2,\ y^2 = u^1 \sin u^2, y^3 = au^2$ are isometric.
    [Hint: Show that the two metrics are the same subject to some conditions.]

Part II

Application of Tensors

7

Application of Tensors in General Theory of Relativity

As mentioned in Chapter 1, the knowledge of the geometry of space is important for Newtonian (classical) as well as Einstein's relativistic mechanics which is reflected in the statement "Dynamics deals with the geometry of motion." To develop geometry of a space, the paramount importance is the assumption of coordinate systems to suitably describe the space concerned based on the corresponding metric. Deformation is an essential notion to invite the concept of "tensors" in non-isotropic medium from an applicable point of view in mathematical science. Tensors being independent of any coordinate system possess the intrinsic property of the geometry of space.

7.1 Introduction

The general theory of relativity is known as the theory of gravitation. For applicability of principle of relativity [5, p. 17] to preserve fully the privileged position among all conceivable frames of reference, the concept of special theory of relativity (STR) based on Lorentz transformation [5, p. 39] is developed. To develop the new theory of gravitation, the idea of privileged position, namely, inertial frames is destroyed to include the most general form of transformation applicable to any positive integral number of dimensions.

Setting aside the Galileo's view of "law of inertia" and the amount of gravitational action of one mass point on another (great mass) of Newton (Kepler), Einstein gave a different interpretation. He has concluded that gravity is not a force (as Newton believed), but the curvature of space–time, and the matter is the source for it and material objects create the gravitational field, which distorts (deforms) or curves the surrounding space–time as the **magnet sets up the magnetic field**. So, the generating space due to the presence of material objects demands the use of non-Euclidean geometry of curved space for its true description. Otherwise, curvilinear coordinates are essential at large as rectilinear coordinates cannot be set up in the curved region of space–time. Hence, curvature of the space–time continuum is the fundamental ingredient for characteristic representation of gravitational theory or the general theory of relativity. Eventually, tensors applicable to all coordinate systems developed in Riemannian space of n dimensions (a manifold) are the best tool (as detected by Einstein) to study general theory of relativity [5,6]. There is no gravity and no curvature, so free particles follow geodesics (shortest path), i.e., straight lines, when the space–time is flat in special theory of relativity.

Therefore, to enter into the threshold of general theory of relativity, it is essential to know the curvature of the space–time, a four-dimensional manifold which is a clear deviation from flat space of special theory of relativity.

7.2 Curvature of a Riemannian Space

Riemann adopted the Gaussian curvature (Section 6.11) of a geodesic surface S at a point P determined by the orientation of the unit vectors $\vec{p}(p^i)$ and $\vec{q}(q^i)$ at P as the definition of (Riemannian) curvature of V_n at that point. The pencil of directions can be expressed as

$$y^i = \left(\alpha p^i + \beta q^i\right)s \tag{7.2.1}$$

with unit tangent $t^i = \left(\dfrac{dy^i}{ds}\right)$ of geodesics at P, where α and β given by $\alpha\, s = u^1$ and $\beta\, s = u^2$ are the current coordinates of points on S generated through

$$y^i = u^1 p^i + u^2 q^i. \tag{7.2.2}$$

If the metric of V_n is denoted by $g_{ij}\, dy^i\, dy^j$ and that of the surface S by $a_{\alpha\beta} du^\alpha du^\beta\left(\alpha,\beta = 1,2\right)$, then

$$a_{\alpha\beta} = g_{ij}\frac{dy^i}{du^\alpha}\frac{dy^j}{du^\beta}. \tag{7.2.3}$$

∴ Using Equation (5.8.5), we can write

$$\left(\Gamma_{ij,k}\right)_g = a_{\beta c}\left[\frac{\partial^2 y^\beta}{\partial x^i \partial x^j} + \left(\Gamma^\beta_{\alpha\gamma}\right)_a \frac{\partial y^\alpha}{\partial x^i}\frac{\partial y^\gamma}{\partial x^j}\right]\frac{\partial y^c}{\partial x^k}$$

$$= a_{\beta c}\frac{\partial y^c}{\partial u^k}\frac{\partial^2 y^\beta}{\partial u^i \partial u^j} + \left(\Gamma_{\alpha\gamma,c}\right)_a \frac{\partial y^\alpha}{\partial u^i}\frac{\partial y^\gamma}{\partial u^j}\frac{\partial y^c}{\partial u^k}$$

Changing variables of the present systems; in this assumption $\Gamma_{\alpha\gamma,c}$ for $a_{\alpha\beta}\,(u^\alpha)$ and $\Gamma_{ij,k}$ for $g_{ij}(y^i)$.

$\therefore \left(\Gamma_{ij,k}\right)_g = \left(\Gamma_{\alpha\gamma,c}\right)_a \dfrac{\partial y^\alpha}{\partial u^i}\dfrac{\partial y^\gamma}{\partial u^j}\dfrac{\partial y^c}{\partial u^k}$, using (7.2.2) where p^i, q^i are constants.

Hence,

$$\left(\overline{|\alpha\beta,\gamma}\right)_g = g_{kl}\frac{\partial y^i}{\partial u^\alpha}\frac{\partial y^j}{\partial u^\beta}\frac{\partial y^k}{\partial y^\gamma}\left(\overline{|}^{\,l}_{ij}\right)_a \tag{7.2.4}$$

$$i \leftrightarrow \alpha$$

$$\text{changing}\quad \begin{matrix} \gamma & \rightarrow & \beta \leftrightarrow j \\ c & \rightarrow & \gamma \leftrightarrow k \end{matrix} \quad \text{on both sides.}$$

But y^i's are the Riemannian coordinates of geodesics of V_n for the metric $g_{ij}\,dy^i\,dy^j$ and Christoffel symbol $\left(\overline{|\alpha\beta,\gamma}\right)_g = 0$.

$\therefore \left(\overline{|}^{\,l}_{ij}\right)_a = 0$ at the origin P, from (7.2.4).

Hence, the Riemannian curvature tensor $R_{h\alpha\beta\gamma}$ for the **surface** S can take nonzero values R_{1212} for two values $\alpha,\beta = 1,2$ subject to the properties $R_{h\alpha\beta\gamma} = -R_{\alpha h\beta\gamma}, R_{h\alpha\beta\gamma} = -R_{h\alpha\gamma\beta}, R_{h\alpha\beta\gamma} = R_{\beta\gamma h\alpha} \because R_{hh\beta\gamma} = R_{h\alpha\beta\beta} = 0$.

$\therefore$ The nonvanishing Riemannian curvature tensors are

$$R_{1212}, -R_{2112}, -R_{1221}, \text{ and } R_{2121}$$

$\because$ In two-dimensional space, the number of independent components is $\frac{1}{12}\left[N^2(N^2-1)\right] = 1$ only, namely, R_{1212}.

$\therefore$ From transformation laws of tensors,

$$\begin{aligned} R'_{1212} &= R_{h\alpha\beta\gamma}\frac{\partial u^h}{\partial u'^1}\frac{\partial u^\alpha}{\partial u'^2}\frac{\partial u^\beta}{\partial u'^1}\frac{\partial u^\gamma}{\partial u'^2} \\ &= R_{1212}\left(\frac{\partial u^1}{\partial u'^1}\right)^2\left(\frac{\partial u^2}{\partial u'^2}\right)^2 - R_{1212}\frac{\partial u^2}{\partial u'^1}\frac{\partial u^1}{\partial u'^2}\frac{\partial u^1}{\partial u'^1}\frac{\partial u^2}{\partial u'^2} \\ &\quad - R_{1221}\frac{\partial u^1}{\partial u'^1}\frac{\partial u^2}{\partial u'^2}\frac{\partial u^2}{\partial u'^1}\frac{\partial u^1}{\partial u'^2} \\ &\quad + R_{2121}\frac{\partial u^2}{\partial u'^1}\frac{\partial u^1}{\partial u'^2}\frac{\partial u^2}{\partial u'^1}\frac{\partial u^1}{\partial u'^2} \end{aligned} \tag{7.2.5}$$

$$\therefore R'_{1212} = R_{1212}\left(\frac{\partial u^1}{\partial u'^1}\frac{\partial u^2}{\partial u'^2} - \frac{\partial u^1}{\partial u'^2}\frac{\partial u^2}{\partial u'^1}\right)^2.$$

The Gaussian curvature κ (Section 6.11) which is invariant for coordinate transformations is defined as

$$\kappa = \frac{R_{1212}}{a} = \frac{R'_{1212}}{a'} \text{ where } a = |a_{\alpha\beta}| = \begin{vmatrix} a_{11} & a_{12} \\ a_{21} & a_{22} \end{vmatrix}$$

$$= \frac{R_{1212}}{a} = \frac{R'_{1212}}{a\left(a_{11}a_{22} - a_{12}^2\right)} \quad \because a' = aJ^2,$$

where $J = \dfrac{\partial u}{\partial u'}$ and $a = |a_{\alpha\beta}|$, $a' = |a'_{\alpha\beta}|$.

Now, by definition,

$$R_{hijk} = g_{ha}R^a_{ijk} = g_{ha}\left[\frac{\partial}{\partial x^j}\left(\Gamma^a_{ik}\right) - \frac{\partial}{\partial x^k}\left(\Gamma^a_{ij}\right) + \Gamma^a_{bj}\Gamma^b_{ik} - \Gamma^a_{bk}\Gamma^b_{ij}\right]$$

$$\therefore R_{1212} = \frac{\partial}{\partial u^1}\left(\Gamma_{22,1}\right)_a - \frac{\partial}{\partial u^2}\left(\Gamma_{21,1}\right)_a. \tag{7.2.6}$$

$\because \left(\Gamma_{\alpha\beta,\gamma}\right)_a = 0$ at the origin P for Riemannian coordinates.

If L, M, N are the second-order and E, F, G are the first-order magnitudes of Gaussian surface* characterized by $ds^2 = Ldu^2 + 2Mdudv + Ndv^2$ and $ds^2 = Edu^2 + 2Fdudv + Gdv^2$, respectively, then

$$\frac{R_{1212}}{a} = \frac{LN - M^2}{EG - F^2}$$

$$a = EG - F^2$$

$$= a_{11}a_{22} - a_{12}^2,$$

where $E = a_{11}$, $F = a_{12} = a_{21}$, $G = a_{22}$.

But $a^{11} = \dfrac{\text{cofactor of } a_{11} \text{ in } |a_{\alpha\beta}| = a}{a} = \dfrac{a_{22}}{a}$.

Similarly, $a^{12} = -\dfrac{a_{12}}{a}$, $a^{22} = \dfrac{a_{11}}{a}$

and

$$a_{11} = g_{ij}\frac{\partial y^i}{\partial u^1}\frac{\partial y^j}{\partial u^1} = g_{ij}p^ip^j = g_{hj}p^hp^j$$

$$a_{22} = g_{ij}\frac{\partial y^i}{\partial u^2}\frac{\partial y^j}{\partial u^2} = g_{ij}q^iq^j = g_{ik}q^iq^k$$

$$a_{12} = g_{ij}\frac{\partial y^i}{\partial u^1}\frac{\partial y^j}{\partial u^2} = g_{ij}p^iq^j = g_{hk}q^kp^h = g_{ji}p^jq^i \tag{7.2.7}$$

$$\therefore a = a_{11}a_{22} - a_{12}^2 = g_{hj}g_{ik}p^hp^jq^iq^k - g_{hk}g_{ij}p^hp^jq^iq^k$$

$$= p^hp^jq^iq^k(g_{hj}g_{ik} - g_{hk}g_{ij}).$$

* Willmore [4].

Now, from (7.2.4),

$$\left(\Gamma_{22,1}\right)_a = g_{hl}\frac{\partial y^i}{\partial u^2}\frac{\partial y^j}{\partial u^2}\frac{\partial y^h}{\partial u^1}\left(\Gamma^l_{ij}\right)_g \quad (k \to h)$$

$$= g_{hl}q^i q^k p^h \left(\Gamma^l_{ik}\right)_g \quad (j \to k)$$

$$\therefore \frac{\partial}{\partial u^1}\left(\Gamma_{22,1}\right)_a = g_{hl}q^i q^k p^h \frac{\partial}{\partial y^j}\left(\Gamma^l_{ik}\right)_g \frac{\partial y^j}{\partial u^1} \tag{7.2.8}$$

$$= g_{hl}q^i q^k p^h p^j \frac{\partial}{\partial y^j}\left(\Gamma^l_{ik}\right)_g.$$

Also,

$$\left(\Gamma_{21,1}\right)_a = g_{hl}\frac{\partial y^i}{\partial u^2}\frac{\partial y^j}{\partial u^1}\frac{\partial y^h}{\partial u^1}\left(\Gamma^l_{ij}\right)_g$$

$$= g_{hl}q^i p^j p^h \left(\Gamma^l_{ij}\right)_g$$

$$\therefore \frac{\partial}{\partial u^2}\left(\Gamma_{21,1}\right)_a = g_{hl}q^i p^j p^h \frac{\partial}{\partial y^k}\left(\Gamma^l_{ij}\right)_g \frac{\partial y^k}{\partial u^2} \tag{7.2.9}$$

$$= g_{hl}q^i p^j p^h q^k \frac{\partial}{\partial y^k}\left(\Gamma^l_{ij}\right)_g.$$

Hence, (7.2.6) can be written as (using 7.2.8 and 7.2.9)

$$R_{1212} = g_{hl}p^h p^j q^i q^k \left[\frac{\partial}{\partial y^j}\left(\Gamma^l_{ik}\right) - \frac{\partial}{\partial y^k}\left(\Gamma^l_{ij}\right)\right]$$

$$= p^h p^j q^i q^k g_{hl} R^l_{ijk} = p^h p^j q^i q^k R_{hijk}.$$

$\therefore$ The curvature

$$\kappa = \frac{R_{1212}}{a} = \frac{p^h p^j q^i q^k R_{hijk}}{p^h p^j q^i q^k \left(g_{hj}g_{ik} - g_{hk}g_{ij}\right)} \tag{7.3.10}$$

is invariant.

This is the mathematical expression of curvature of the Riemannian space V_n with R_{hijk}. This ascertains the nomenclature "covariant curvature tensor" for R_{hijk}. Hence, R_{hijk} (or $R^h{}_{ijk}$) characterizes the behavioral properties of the curve space of general theory of relativity.

7.3 Flat Space and Condition for Flat Space

Definition

If the curvature κ of the Riemannian space V_n, namely R_{hijk} vanishes, it is called flat space.

Condition for Flat Space

From the definition of curvature shown in (7.3.10), where R_{hijk} is the covariant curvature tensor, and p^i, q^j are the unit vectors showing orientation at the origin P of geodesics.

$$\kappa = \frac{p^h p^j q^i q^k R_{hijk}}{p^h p^j q^i q^k (g_{hj} g_{ik} - g_{ij} g_{hk})}.$$

For flat space $\kappa = 0 \Longrightarrow R_{hijk} = 0$, the bracket in the denominator must not be zero.

$\therefore$ If $R_{hijk}\left(= g_{ha} R^a_{ijk}\right) = 0$, or if all the components of R_{hijk} or R^a_{ijk} are zero, then the curvature will be zero, and the space will be flat. This is the required condition for flat space.

But $R_{hijk} = \frac{1}{2}\left[\frac{\partial^2 g_{ij}}{\partial x^h \partial x^k} + \frac{\partial^2 g_{hk}}{\partial x^i \partial x^j} - \frac{\partial^2 g_{ik}}{\partial x^h \partial x^j} - \frac{\partial^2 g_{hj}}{\partial x^i \partial x^k}\right] + g_{ab} \Gamma^a_{hk} \Gamma^b_{ij} - g_{ab} \Gamma^a_{hj} \Gamma^b_{ik}.$

If g_{ij}'s are constants, $\Gamma_{ij,k} = \Gamma^b_{ij} = 0, \therefore R_{hijk} = 0.$

Hence, g_{ij} = constants are basically the conditions for flat space.

Note: STR is restricted to flat space only.

7.4 Covariant Differential of a Vector

Let A^i be the components of a vector at the point x^i and $A^i + dA^i$ be the components at a neighboring point $x^i + dx^i$ in a vector field. Therefore, the difference dA^i of the two vectors $A^i + dA^i$ and A^i being ordinary differential is a vector.

The transformation of the coordinates gives $dx'^i = \frac{\partial x'^i}{\partial x^k} dx^k$, otherwise, $A'^i = \frac{\partial x'^i}{\partial x^k} A^k$ stands for the transformation of a vector.

But

$$dA'^i = \frac{\partial x'^i}{\partial x^k} dA^k + \frac{\partial^2 x'^i}{\partial x^j \partial x^k} dx^j A^k, \tag{7.4.1}$$

which is not a vector for the presence of the second term. Of course, for linear transformation belonging to rectilinear coordinate systems, this transformation characteristically behaves like a vector since the second term vanishes in that case. On the other hand, the difference of two vectors needs to be a vector in a general coordinate system. Hence, to get this difference to be a vector in a curvilinear coordinate system, it necessitates to translate a vector A^i at x^i to the location of $x^i + dx^i$ of the other vector $A^i + dA^i$ so that they are located at the **same point**. This translation is related to **parallel translation to itself**. In a general curvilinear coordinate system, this translation by itself changes the components of the vector, and the changes are denoted by δA^i different from ordinary differential dA^i. From (7.4.1), it is observed that this change δA^i should be proportionate to both the vectors A^k (i.e., A^i) and the displacement $dx^j \left(\text{if} \dfrac{\partial^2 x'^i}{\partial x^j \partial x^k} \neq 0 \right)$.

Hence, the difference between the original vector $A^i + dA^i$ and the transported vector $A^i + \delta A^i$ at the point $(x^i + dx^i)$ is given by

$$\left(DA^i\right) = dA^i - \delta Ai. \tag{7.4.2}$$

$\therefore$ In consistent with the above changes in the components A^i of the vector due to translation and the components of the displacement dx^i, the changes δA^i in (7.4.2) can be written as†

$$\delta A^i = -\Gamma^i_{jk} A^k dx^j. \tag{7.4.3}$$

(Γ^i_{jk} are some functions of coordinates x^i)

Thus, (7.4.2) is called the **covariant differential** of the given vector A^i.

7.5 Motion of Free Particle in a Curvilinear Co-Ordinate System for Curved Space

If $u^i = \dfrac{dx^i}{ds}$ ($i = 1, 2, 3, 4$) are the four vectors (velocity) tangential to the time track of a free particle, it is represented in a rectilinear coordinate system by the equations $du^i = 0$.

This gives $x^i = As + B$ which is the intrinsic equation of a straight line in a four-dimensional continuum. Eventually, $du^i = 0$ characterizes the uniform rectilinear motion in three-dimensional physical space also. But to write the **corresponding** equations of motion in a curvilinear coordinate system, we

† Section 6.4 of Ref. [5].

are to make use of **covariant differential** instead of ordinary differential. Hence, the equations of motion of free particle in the curvilinear coordinate system are given by

$$\left(Du^i =\right)du^i - \delta u^i = 0 \tag{7.5.1}$$

i.e., $du^i + \Gamma^i_{jk} u^k dx^j = 0$ from Equations (7.4.2) and (7.4.3),

$$\text{i.e.,}\ \frac{d^2x^i}{ds^2} + \Gamma^i_{jk}\frac{dx^j}{ds}\frac{dx^k}{ds} = 0 \tag{7.5.2}$$

which are the differential equations of geodesics, and contextually, they characterize the world line (track) of **free particle** in the general coordinate system.

If we compare Equation (7.5.2) or $\frac{d^2x^i}{ds^2} = -\Gamma^i_{jk}\frac{dx^j}{ds}\frac{dx^k}{ds}$ with Newtonian gravitational equation with potential ϕ, namely, $\frac{d^2x^i}{dt^2} = -\delta^{ij}\frac{d\phi}{dx^j}$, then Γ^i_{jk} can be attributed to "forces" that arise inherently in the system. Hence, $\Gamma^i_{jk} \neq 0$ or $g_{ij} \neq$ constant is the **generating factor of the potential** (ϕ) and responsible for the curvature of the path of free particle caused by gravitational field or mass energy of matter in curved space.

7.6 Necessity of Ricci Tensor in Einstein's Gravitational Field Equation

In classical (Newtonian) mechanics, the field equation in the presence of matter according to Newton's law of gravitation is described by Poisson's equation:

$$\begin{aligned} &\nabla^2\phi = 4\pi\gamma\rho, \\ &\text{i.e.,}\ \frac{\partial^2\phi}{\partial x^2} + \frac{\partial^2\phi}{\partial y^2} + \frac{\partial^2\phi}{\partial z^2} = 4\pi\gamma\rho, \end{aligned} \tag{7.6.1}$$

where ϕ is the gravitational potential, ρ is the density of distributed matter, and γ is the gravitational constant. So, to get its general relativistic analogue, we need to correlate these quantities suitably under physical system introducing general relativistic concepts.

1. It is discussed is Section 7.1 that the material energy creates the gravitational field to turn the space–time continuum into a curved space. Therefore, the counterpart of Equation (7.6.1) should be made applicable to **curved space**.
2. For the "principle of covariance,"‡ an essence of general relativity, all natural laws must be expressed in **tensor forms** (Covariant) for their validity in all coordinate systems including non-inertial frames of curved space required for general theory of relativity. Hence, Equation (7.6.1) needs to be expressed completely in tensor form.
3. It is shown in Section 7.5 that the equations of motions of free fall of a particle in curvilinear coordinates of space–time are characterized by the geodesic equations $\frac{d^2x^i}{ds^2}+\Gamma^i_{jk}\frac{dx^j}{ds}\frac{dx^k}{ds}=0$, which is reducible to Newton's equation of motions $\frac{d^2x^i}{dt^2}=-\frac{\partial\phi}{dx^i}(=-\nabla\phi)$, where ϕ is the gravitational potential.
   It has already been mentioned in Section 7.5 that g_{ij} ($\neq$ constant) is responsible to generate gravitational potential $\phi\left[g_{44}=1\pm\frac{2\phi}{c^2}\right]$ [Section 7.3 of Ref. (5, p. 191)]. Hence, in relativistic theory of gravitation, ϕ is to be replaced by the metric tensor $g_{\mu\upsilon}$ ($\neq$ constant) or by some relation with $g_{\mu\upsilon}$. Moreover, the left-hand side (L.H.S.) of Equation (7.6.1) does not involve the derivatives of ϕ higher than two (second), so the replacement tensor (for covariant form) must also contain the second-order derivatives of $g_{\mu\upsilon}$, but the Ricci tensor $R_{\mu\upsilon}$ is the tensor with second-order derivatives of $g_{\mu\upsilon}$ as
   $$R_{\mu\upsilon}=\frac{\partial}{\partial x^{\upsilon}}\left(\Gamma^{\alpha}_{\mu\alpha}\right)-\frac{\partial}{\partial x^{\alpha}}\left(\Gamma^{\alpha}_{\mu\upsilon}\right)+\Gamma^{\alpha}_{\beta\upsilon}\Gamma^{\beta}_{\mu\alpha}-\Gamma^{\alpha}_{\beta\alpha}\Gamma^{\beta}_{\mu\upsilon}$$
   $$\text{with}\ \Gamma^{\alpha}_{\mu\upsilon}=g^{\alpha a}\Gamma_{\mu\upsilon,a}=\frac{1}{2}g^{\alpha a}\left[\frac{\partial g_{\mu a}}{\partial x^{\upsilon}}+\frac{\partial g_{\upsilon a}}{\partial x^{\mu}}-\frac{\partial g_{\mu\upsilon}}{\partial x^{a}}\right]$$
   $$\text{and}\ \frac{\partial}{\partial x^{\upsilon}}\left[\Gamma^{\alpha}_{\mu\alpha}\right]=\frac{\partial^2}{\partial x^{\upsilon}\partial x^{\mu}}(\log\sqrt{\pm g}).$$
   Hence, the suitable relativistic analogue of the L.H.S. of Equation (7.6.1) is the Ricci tensor, if it satisfies **some other** conditions in conformity with the right-hand side (R.H.S.).
4. In agreement with the L.H.S. of (7.6.1), we need a second-order tensor for the R.H.S. in place of density ρ of matter.
   The gravitational field is the outcome (effect) of mass distribution or mass–energy distribution of matter, and energy momentum tensor $T_{\mu\upsilon}$ or $T^{\mu\upsilon}$ characterizes the cause of the distributions. Hence, the density ρ of material particles generating the force of gravity is

‡ Section 6.1 of Ref. [5, p. 155].

to be represented by the second-order "energy momentum" tensor $T^{\mu\upsilon}$>(or $T_{\mu\upsilon}$).

Also, the energy momentum tensor ($T^{\mu\upsilon}$ or $T_{\mu\upsilon}$) of a closed system (like one comprising the material distribution and the force, or the field together) requires to be bounded by the conservation laws

$T^{\mu}_{\upsilon,\mu} = 0$ or $\left(g^{\alpha\upsilon}T^{\mu}_{\upsilon}\right)_{,\mu} = 0$ or $T^{\alpha\mu}_{,\mu} = 0$,

which is divergence free and $T^{\alpha\mu} = T^{\mu\alpha}$, i.e., symmetric.

5. In light of conservation laws $T^{\mu}_{\upsilon,\mu} = 0$ as satisfied by the energy momentum tensor, it is desirable to make it applicable to the Ricci tensor $R_{\mu\upsilon}$ also. But the Einstein tensor $G^{\mu}_{\upsilon} = \left(R^{\mu}_{\upsilon} - \frac{1}{2}\delta^{\mu}_{\upsilon}R\right)$ in terms of Ricci tensor is divergence free, i.e., $G^{\mu}_{\upsilon,\mu} = \left(R^{\mu}_{\upsilon} - \frac{1}{2}\delta^{\mu}_{\upsilon}R\right)_{,\mu} = 0.$

   Otherwise, $\left(g^{\alpha\upsilon}R^{\mu}_{\upsilon} - \frac{1}{2}g^{\alpha\upsilon}\delta^{\mu}_{\upsilon}R\right)_{,\mu} = 0$ or $\left(R^{\alpha\mu} - \frac{1}{2}g^{\alpha\mu}R\right)_{,\mu} = 0.$

   Also, $R^{\mu\alpha} = R^{\alpha\mu}$, which is symmetric.

   Hence, the Ricci tensor $R^{\mu\alpha}$ (or $R_{\mu\upsilon}$) also satisfies the required other conditions like $T^{\mu\upsilon}$.

   ∴ Based on the aforesaid requirements (1)–(5), Einstein adopted the field equations for general theory of relativity as $R^{\mu\upsilon} - \frac{1}{2}g^{\mu\upsilon}R = -KT^{\mu\upsilon}$ or $R_{\mu\upsilon} - \frac{1}{2}g_{\mu\upsilon}R = -KT_{\mu\upsilon}$.

These are the **essential field equations for general theory of relativity,** which were expanded subsequently after various cosmological considerations. This is purely a subject matter of "general theory of relativity," and it is beyond the scope of the book, to discuss it completely.

Of course, in the absence of matter (or gravitational field),

$$T^{\mu\upsilon} = T^{\mu\upsilon} = 0,\ g_{\mu\upsilon} = \text{constant}.$$

The field equations reduce to $R^{\mu\upsilon} = 0 = R_{\mu\upsilon}$ **for empty space** which is identical to $\nabla^2\phi = 0$, which is Newton's vacuum equation.

N.B.: The importance of Ricci tensor in general theory of relativity is beyond description. In this pursuit, it is amplified only the use of tensors (primarily the Ricci tensor) in general theory of relativity.

8

Tensors in Continuum Mechanics

8.1 Continuum Concept

In the investigation of material behavior of a body or medium, the bulk of the matter is considered as a whole but not the individual molecule. For this reason, the observed macroscopic behavior is counted in general by assuming that the material is continuously distributed throughout its volume and completely fills the space it occupies instead of considering its molecular distributions. This **continuum concept** of matter is the fundamental postulate of continuum mechanics.

8.2 Mathematical Tools Required for Continuum Mechanics

The physical quantities related to continuum mechanics for its in-depth description are independent of any particular coordinate system for reference. Generally, from a mathematical point of view, these physical quantities conveniently need to describe by means of referring them to some coordinate system. Eventually, tensors which are independent of any particular coordinate system are the appropriate tools to adopt in this consideration. Hence, physical laws of continuum mechanics are expressed in terms of tensor equations. Usually, tensor transformations are linear and homogeneous, and if they (laws of continuum mechanics) are expressed in the form of tensor equations in one coordinate system, they remain valid in any other coordinate system. This invariance of tensor equations under coordinate transformations is one of the principal reasons (similar to general theory of relativity) for the utility of tensors in continuum mechanics. Of course, **Cartesian tensors** are sufficient to deal with continuum mechanics, and hence, it can be developed with reference to Cartesian tensors.

In Euclidean space of three dimensions, the number of components of a tensor of order n is 3^n. A vector is a tensor of order one with 3^1 components.

The **stress and strain** are the second-order tensors having $3^2=9$ components in general. The physical quantities (mentioned earlier) involved in the study of continuum mechanics are the stresses and strains, which are invariably related to the deformation of media and bodies. Second-order tensors are also known as "dyadics," and the quantities in continuum mechanics are represented by dyadics [11].

Definition

Stress: The forces occurring in a bulk of the material medium proportional to the mass of the substance (e.g., gravity, magnetic force, centrifugal force) are known as body forces, which are measured per unit volume.

The forces acting over the volume of bounding surface of a body and measured in units of force per unit area are called surface forces. This force per unit area is called **stress** which gives the measure of the **intensity** of the reaction of the material lying on one side of the material or that which lies on the other side. For this inherent property, the tendency of deformation of the state of a body or medium is bound to be acted by stress. Stress being a force per unit area must have **two directions** those of the force and normal to the area associated with it.

The stress tensor components perpendicular to the surface (or plane face) are called normal stress (σ) and tangential to the surface (or plane face) are called shearing stress (τ).

Definition

Strain: The deformation caused by stress which may be dilated resulting in change in volume or distortion with changes in form or both is called **strain**. Otherwise, a distortion, deformation of change in the position of particles relative to each other, is known as strain.

The change in confining pressure can change in volume but in shape for isotropic bodies where mechanical properties are uniform in all directions. With increasing/decreasing confining pressure, the volume of the body decreases/increases, and dilation is negative/positive accordingly.

Mathematically, if Δf_i (f_i body force) is the resultant force exerted across the surface element Δs of S enclosing a volume V, then the Cauchy's stress principle states that the average **force per unit area** on Δs, namely, $\frac{\Delta f_i}{\Delta s}$, tends to a finite limit $\frac{df_i}{ds}$ as $\Delta s \to 0$ at a point P of the surface. Symbolically, the stress vector is written as $\vec{t}_{(\hat{n})i} = \lim_{\Delta s \to 0} \frac{\Delta f_i}{\Delta s} = \frac{\overline{df}_i}{ds}$ when the moment of Δf_i at the point P vanishes in the limiting process. The stress vectors **are different at the same time** for different surfaces containing the same point P. The stress principle is necessary to know the state of stress at a point in a medium in motion or a body subjected to deformation.

8.3 Stress at a Point and the Stress Tensor

In a continuum, the Cauchy's stress principle associates a stress vector $\vec{t}_{(\hat{n}_i)}$ with each unit normal $\hat{n}_i$ representing the orientation of an infinitesimal surface element having an arbitrary internal point P.

The **totality** of all possible pairs of such combined vectors $\vec{t}_{(\hat{n}_i)}$ and $\hat{n}$ at P defines the state of stress at that point. Every pair of stresses and normal vectors are essential to represent the state of stress at that point. Eventually, the stress vectors on each of the three mutually perpendicular planes (for Cartesian system) through P can give the state of stress. Of course, coordinate transformation equations can serve to relate the stress vector on any other plane at that point with the given three planes. The state of stress is perfectly homogeneous under the application of a force **without rotation**.

The stress vectors in the coordinate plane surfaces of a cubic element with mutually perpendicular axes $1(X_1)$, $2(X_2)$, $3(X_3)$ can be written as

$$\begin{aligned}\vec{t}_{(\hat{n}_1)} &= t_{(\hat{n}_1)_1}\hat{n}_1 + \vec{t}_{(\hat{n}_1)_2}\hat{n}_2 + t_{(\hat{n}_1)_3}\hat{n}_3 \\ &= t_{(\hat{n}_1)_j}\hat{n}_j\end{aligned}$$

Similarly, $\vec{t}_{(\hat{n}_2)} = t_{(\hat{n}_2)_j}\hat{n}_j$ and $\vec{t}_{(\hat{n}_3)} = t_{(\hat{n}_3)_j}\hat{n}_j$

The nine components are $t_{(\hat{n}_i)_j} \equiv \sigma_{ij}$ or τ_{ij} which is called the second-order (Cartesian) **stress tensor**. Here, $t_{(\hat{n}_i)}$ is the stress vector in the direction $\hat{n}_i$ perpendicular to the surface, and $t_{(\hat{n}_i)_j}$ is its resolved part in the jth direction.

The stress tensor can be expressed in matrix form as

$$\sigma_{ij} = \begin{pmatrix} \sigma_{11} & \sigma_{12} & \sigma_{13} \\ \sigma_{21} & \sigma_{22} & \sigma_{23} \\ \sigma_{31} & \sigma_{32} & \sigma_{33} \end{pmatrix} \text{or} \begin{pmatrix} \sigma_{xx} & \sigma_{xy} & \sigma_{xz} \\ \sigma_{yx} & \sigma_{yy} & \sigma_{yz} \\ \sigma_{zx} & \sigma_{zy} & \sigma_{zz} \end{pmatrix}$$

with reference to the coordinate planes (shown in Figure 8.1). The components in the diagonal $\sigma_{11}=\sigma_{xx}$, $\sigma_{22}=\sigma_{yy}$, $\sigma_{33}=\sigma_{zz}$ are called normal stresses perpendicular to planes, and the components σ_{ij} $(i \neq j)$ or σ_{xy}, σ_{yx}, σ_{xz}, σ_{zx}, σ_{yz}, σ_{zy} tangential to the coordinate planes are called shear stresses (τ_{ij}) in other notations.

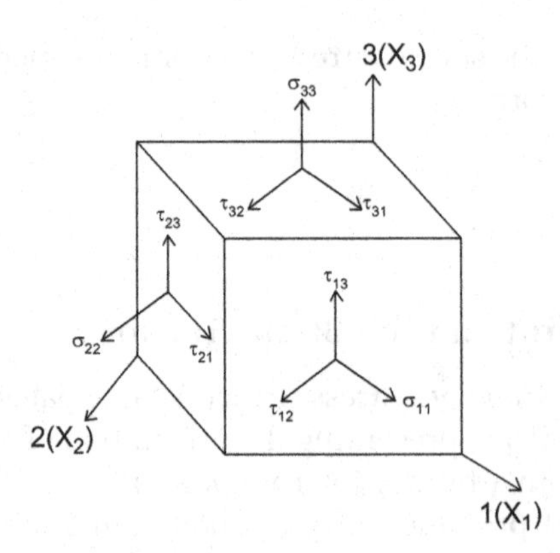

FIGURE 8.1
Stress components on the plane surfaces of a cubic element (for simplicity).

8.4 Deformation and Displacement Gradients

Let (X_1, X_2, X_3) be the material coordinates of the point P_o in undeformed configuration of a material continuum at $t=0$, with respect to $O\text{–}X_1X_2X_3$, and $P(x_1, x_2, x_3)$ be the corresponding position in the deformed configuration at $t=t_1$ with respect to the Cartesian system $O'-x_1x_2x_3$ so that $\left(\overline{OP}_o=\right)\vec{X}=X_1\hat{I}_1+X_2\hat{I}_2+X_3\hat{I}_3$ and $\left(\overline{O'P}=\right)\vec{x}=x_1\hat{e}_1+x_2\hat{e}_2+x_3\hat{e}_3$.

The particles of the continuum undergoing deformation can move along different paths in space. If the particle P_o initially at $t=0$ is assumed to move to the position P at $t=t_1$, then functionally, it can be represented by $x_i=x_i(X_1, X_2, X_3, t)=x_i\,(X, t)$ or conversely $X_i=X_i(x_1, x_2, x_3, t)=X_i\,(x, t)$.

$$\therefore dx_i=\frac{\partial x_i}{\partial X_j}dX^j \tag{8.4.1}$$

where the tensor $\frac{\partial x_i}{\partial X_j}$ is called the **material deformation gradient tensor,** and t is the absolute time. Similarly, from $X_i=X_i(x, t)$,

$$dX_i=\frac{\partial X_i}{\partial x_j}dx^j \tag{8.4.2}$$

It can be seen that $\frac{\partial X_i}{\partial x_j}$ is also a tensor and is called **spatial deformation gradient tensor**.

Definitely, the material and spatial deformation tensors are interrelated by means of the chain rule of partial differential:

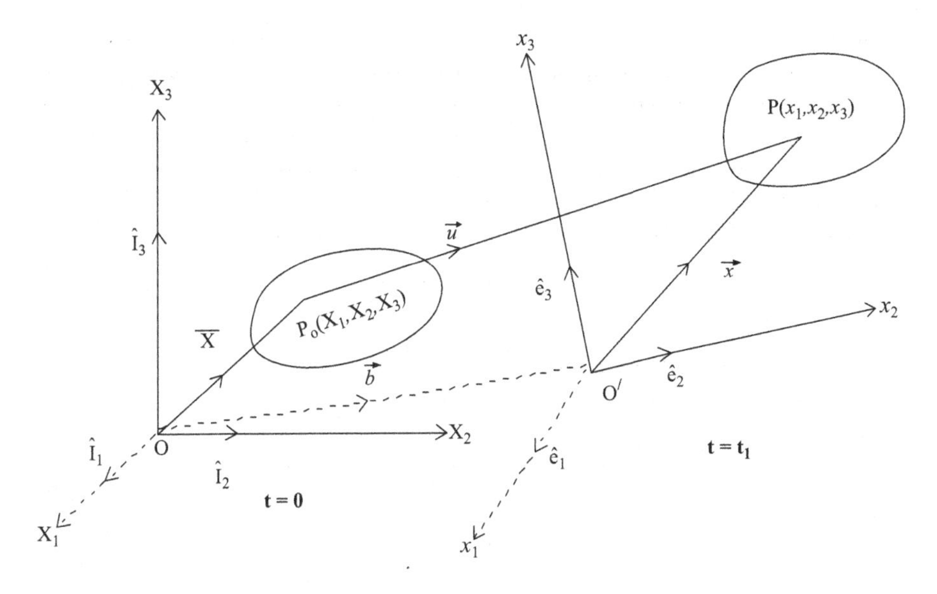

FIGURE 8.2
Deformation graph.

$$\frac{\partial x_i}{\partial X_j}\frac{\partial X_j}{\partial x^k} = \delta_{ik}\frac{\partial X_i}{\partial x^j}\frac{\partial x^j}{\partial X_k} \tag{8.4.3}$$

From the displacement vector $u_i = x_i - X_i + b_i$ (Figure 8.2), the partial differentiation with respect to coordinates gives the material displacement gradient $\frac{\partial u_i}{\partial X_j}$ and the material displacement $\frac{\partial u_i}{\partial x_j}$ as

$$\frac{\partial u_i}{\partial X_j} = \frac{\partial x_i}{\partial X_j} - \frac{\partial X_i}{\partial X_j} = \frac{\partial x_i}{\partial X_j} - \delta_{ij} \tag{8.4.4}$$

and

$$\frac{\partial u_i}{\partial x_j} = \frac{\partial x_i}{\partial x_j} - \frac{\partial X_i}{\partial x_j} = \delta_{ij} - \frac{\partial X_i}{\partial x_j} \tag{8.4.5}$$

8.5 Deformation Tensors and Finite Strain Tensors

Let us consider two **superimposed** rectangular Cartesian coordinate systems $O\text{–}X_1X_2X_3$ for initial configuration and $O\text{–}x_1x_2x_3$ for final configuration after deformation of a material continuum. To extract the characteristic difference of the measure of material deformation, it is justified to refer to a different coordinate system (instead of referring to the same system) but from the **same position**. The neighboring particles P_o and Q_o before deformation are supposed to move to the points P and Q, respectively, in the deformed configuration (Figure 8.3).

The square of the differential element of length between P_o and Q_o

$$dX^2 = dX_i dX_i = \delta_{ij} dX_i dX_j \tag{8.5.1}$$

where $dX_i = \frac{\partial X_i}{\partial x_j} dx_j$ (from 8.4.2).

$$\begin{aligned}(dX)^2 = dX_k dX_k &= \frac{\partial X_k}{\partial x_i}\frac{\partial X_k}{\partial x_j} dx_i dx_j \\ &= C_{ij} dx_i dx_j\end{aligned} \tag{8.5.2}$$

in which the second-order tensor $C_{ij} = \frac{\partial X_k}{\partial x^i}\frac{\partial X_k}{\partial x^j}$ is called the **Cauchy's deformation tensor**.

Also, the square of the differential element of length between P and Q for the deformed configuration is

$$(dx)^2 = dx_i dx_i = \delta_{ij} dx_i dx_j, \tag{8.5.3}$$

where $dx_i = \frac{\partial x_i}{\partial X_j} dX_j$ (from 8.4.1).

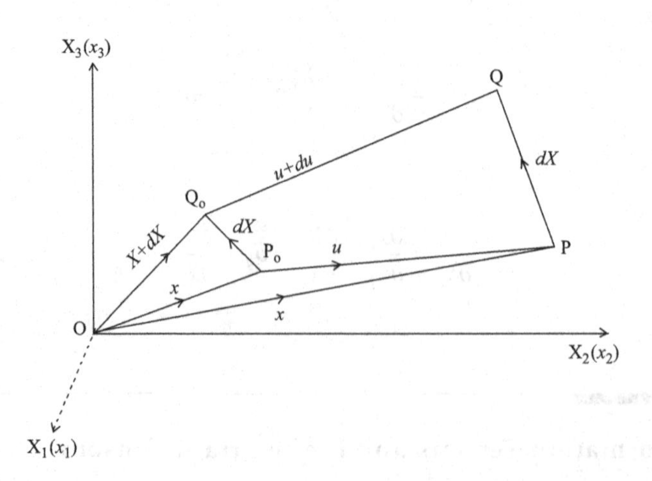

FIGURE 8.3
Graph of deformation tensor and strain tensor.

It can be written as

$$(dx)^2 = \frac{\partial x_k}{\partial X_i}\frac{\partial x_k}{\partial X_j}dX_i dX_j = G_{ij}dX_i dX_j, \tag{8.5.4}$$

where the second-order tensor $G_{ij} = \frac{\partial x_k}{\partial X_i}\frac{\partial x_k}{\partial X_j}$ is called the **Green's deformation tensor**.

The difference $(dx)^2 - (dX)^2$ corresponding to two neighboring particles between the initial and final configurations of a material continuum is taken as the **measure of deformation**. For all neighboring particles, if this difference vanishes for a continuum, the corresponding displacement is called rigid displacement.

Now, $(dx)^2 - (dX)^2 = (G_{ij} - \delta_{ij})dX_i dX_j = 2L_{ij}dX_i dX_j$, using (8.5.4) and (8.5.1) where

$$L_{ij} = \frac{1}{2}\left(G_{ij} - \delta_{ij}\right) = \frac{1}{2}\left(\frac{\partial x_k}{\partial X_i}\frac{\partial x_k}{\partial X^j} - \delta_{ij}\right) \tag{8.5.5}$$

which is called the **Lagrangian (or Green's) finite strain tensor**.

Also, $(dx)^2-(dX)^2=(\delta_{ij}-C_{ij})dx_i dx_j$ using (8.5.3) and (8.5.2)$=2E_{ij}dx_i dx_j$, using (8.5.3) and (8.5.2)

where

$$E_{ij} = \frac{1}{2}\left(\delta_{ij} - C_{ij}\right) = \frac{1}{2}\left(\delta_{ij} - \frac{\partial X_k}{\partial x_i}\frac{\partial X_k}{\partial x_j}\right) \tag{8.5.6}$$

which is called the **Eulerian finite strain tensor**.

Making use of the relation of displacement vector,

$$u_i = x_i - X_i + b_i$$

L_{ij} can be written as

$$\begin{aligned} L_{ij} &= \frac{1}{2}\left[\left(\frac{\partial u_k}{\partial X_i} + \delta_{ki}\right)\left(\frac{\partial u_k}{\partial X_j} + \delta_{kj}\right) - \delta_{ij}\right] \\ &= \frac{1}{2}\left[\frac{\partial u_k}{\partial X_i}\frac{\partial u_k}{\partial X_j} + \frac{\partial u_j}{\partial X_i} + \frac{\partial u_i}{\partial X_j} + \delta_{ij} - \delta_{ij}\right] \\ &= \frac{1}{2}\left[\frac{\partial u_i}{\partial X_j} + \frac{\partial u_j}{\partial X_i} + \frac{\partial u_k}{\partial X_i}\frac{\partial u_k}{\partial X_j}\right] \end{aligned} \tag{8.5.7}$$

Similarly,

$$E_{ij} = \frac{1}{2}\left[\frac{\partial u_i}{\partial x_j} + \frac{\partial u_j}{\partial x_i} - \frac{\partial u_k}{\partial x_i}\frac{\partial u_k}{\partial x_j}\right] \tag{8.5.8}$$

Now, if we impose the condition for small deformation theory on the displacement gradients, which are very small compared to unity, then the product term in (8.5.7) can be ignored. Hence, the expression corresponding to L_{ij} can be written as

$$l_{ij} = \frac{1}{2}\left(\frac{\partial u_i}{\partial X_j} + \frac{\partial u_j}{\partial X_i}\right) \tag{8.5.9}$$

It is called the Lagrangian infinitesimal strain tensor.

Also, for $\frac{\partial u_k}{\partial x_i} \ll 1$, ignoring the product term, (8.5.8) can be equivalently written as

$$e_{ij} = \frac{1}{2}\left(\frac{\partial u_i}{\partial x_j} + \frac{\partial u_j}{\partial x_i}\right) \tag{8.5.10}$$

which is called the **Eulerian infinitesimal strain tensor**.

8.6 Linear Rotation Tensor and Rotation Vector in Relation to Relative Displacement

Let $\vec{u}^{(P_o)}$ and $\vec{u}^{(Q_o)}$ represent the displacement vectors of the two neighboring particles in a material medium.

$$\therefore d\vec{u} = \vec{u}^{(Q_o)} - \vec{u}^{(P_o)}$$

$$\therefore du_i = u_i^{(Q_o)} - u_i^{(P_o)}$$

which is the measure of **relative displacement vector** of the particle originally at Q_o with respect to the particle originally at P_o. Subject to continuous displacement, $u_i^{(P_o)}$ the expansion about by Taylor's series, it can be written as $du_i = \left(\frac{\partial u_i}{\partial X_j}\right)_{P_o} dX_j$ (neglecting higher order terms for small displacement).

The material displacement $\frac{\partial u_i}{\partial X_j}$ can be decomposed into a symmetric part and an antisymmetric part so that

$$du_i = \left[\frac{1}{2}\left(\frac{\partial u_i}{\partial X_j} + \frac{\partial u_j}{\partial X_i}\right) + \frac{1}{2}\left(\frac{\partial u_i}{\partial X_j} - \frac{\partial u_j}{\partial X_i}\right)\right]_{P_o} dX_j \tag{8.6.1}$$

$$\therefore \frac{\partial u_i}{\partial X_j} = l_{ij} + W_{ij} \quad (\text{using } 8.59),$$

where

$$W_{ij} = \frac{1}{2}\left(\frac{\partial u_i}{\partial X_j} - \frac{\partial u_j}{\partial X_i}\right) \tag{8.6.2}$$

which is called the linear Lagrangian rotation tensor (Figure 8.4).

For infinitesimal rigid body rotation, corresponding to relative displacement at point P_o, the Lagrangian strain tensor l_{ij} vanishes. Hence, the infinitesimal rotation vector is thus given by

$$w_i = \frac{1}{2} \in_{ijk} W_{kj} \tag{8.6.3}$$

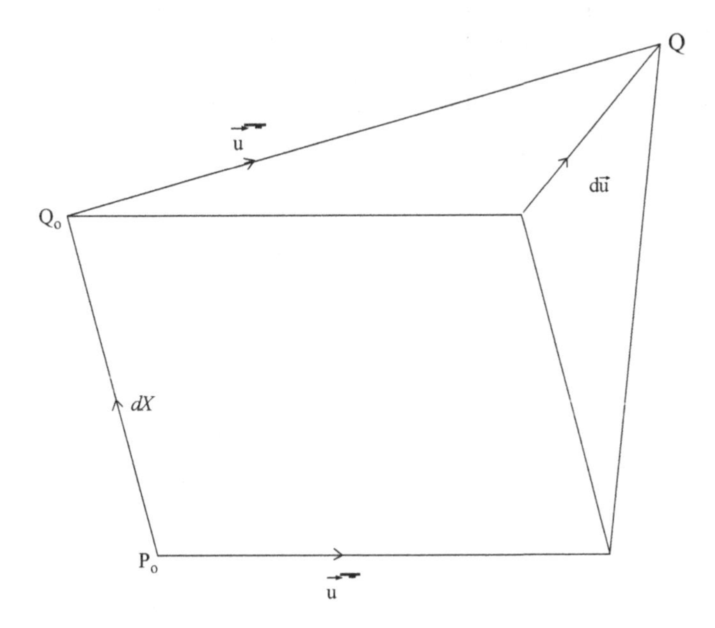

FIGURE 8.4
Graph for rotation tensor and vector.

Again, according to Eulerian description of the relative displacement vector, it can be written as

$$du_i = \frac{\partial u_i}{\partial x_j} dx_j$$

so that

$$\frac{\partial u_i}{\partial x_j} = \left[\frac{1}{2}\left(\frac{\partial u_i}{\partial x_j} + \frac{\partial u_j}{\partial x_i} \right) + \frac{1}{2}\left(\frac{\partial u_i}{\partial x_j} - \frac{\partial u_j}{\partial x_i} \right) \right] \tag{8.6.4}$$

$$= e_{ij} + w_{ij},$$

where

$$w_{ij} = \frac{1}{2}\left(\frac{\partial u_i}{\partial x_j} - \frac{\partial u_j}{\partial x_i} \right) \tag{8.6.5}$$

which is called the **Eulerian rotation tensor** with Eulerian rotation vector

$$w_i = \frac{1}{2} \in_{ijk} w_{kj}. \tag{8.6.6}$$

It has already been discussed that deformation is connected with displacements. Therefore, motion of any continuum such as fluid or gas requires the use of tensors (or vectors), namely, stress–strain tensors, rotation tensor, and displacement vector. Ironically, there is no alternative but to use tensors in non-isotropic medium to arrive at accurate results of an investigation.

9

Tensors in Geology

9.1 Introduction

The structural investigations of spasmodic **deformation** [it means the change in shape of a body from the initial (undeformed) configuration to a subsequent (deformed) configuration] caused by nature and origin of some forces need application of some mathematical concepts such as tensors. The mathematical entities such as stress and strain (Section 8.2) are largely responsible to study any kind of deformation. A second-order tensor called "stress" is the essence of structural geology.

Rheology is closely related to the study of deformation of the structures of the earth and any kind of material structure ranging from the order of seconds (seismic-wave propagation) to hundreds of millions of years (geodynamics). The aforesaid "stress and strain" are the fundamental ingredients to deal with the analysis of continuum mechanics of deformation (discussed in Section 8.2) of extended bodies in the context of rheology. Newtonian and non-Newtonian viscosity, linear rheological bodies, plasticity, and brittle failure can be investigated with these mathematical entities—the second-order tensors. From a rheological standpoint, properties of lithosphere and the mantle, temperature distribution of lithosphere, thermal convection in the mantle, flexure of the lithosphere, stresses on it, and viscosity of the mantle from surface loading data are some of the features within this domain. Of course, atomic basic deformation and flow in polycrystalline materials covering hydrolytic weakening dynamic recrystallization and pressure and temperature effects may attract the involvement of both geologists and geophysicists [Ref. 7].

The displacement gradient tensor (one of many tensors related to strain) relates the **position vector of a point** to the displacement of the point during a displacement. Knowing displacement gradient tensor, it is possible to calculate how all points within a body are displaced as a function of position during deformation. If we know the **stress tensor**, we can calculate the **stress vector on a plane** of any orientation within a body, which is largely important in any study of earthquakes, induced seismicity, faulting, etc.

There are three stages of deformation (mentioned earlier) when direct forces act on a body extending to a short period of time, minutes, or hour. They are as follows:

i. Elastic
ii. Plastic
iii. Rupture

Elastic: If the body returns to the original shape and size after the withdrawal of the stress, the deformation is called **elastic**. If the body does not return to the original shape when the stress exceeds a certain stage, it is called **elastic limit**.

The strain is proportional to stress when it remains always less than the elastic limit, and the deformation obeys Hooke's law.

Plastic: The deformation is said to be plastic if the stress exceeds the elastic limit.

N.B.: The difference between the external force applied to a body and the corresponding outcomes of internal actions and reactions generates stress.

Rupture: If the specimen is subjected to continuous increase of stress, one or more fractures based on several factors can develop, which eventually fails by rupture. Rupture is responsible for "**brittle**" in substances before certain stage of plastic deformation.

From a geometrical point of view, strain causing distortion of a body can also be classified as homogeneous and inhomogeneous.

Homogeneous deformation: After deformation,

i. Straight lines remain straight lines.
ii. Parallel lines remain parallel.
iii. In the strained body, all lines in the same direction have **constant value** of e (extension—change in unit in length), λ (quadratic elongation), ψ (angular strain), and γ (shear strain).

Inhomogeneous deformation: After deformation,

i. Straight lines change to curves.
ii. Parallel lines turn nonparallel.
iii. The values of the above four parameters e, λ, ψ, and γ in any one given direction of the body become variable, not constant.

9.2 Equation for the Determination of Shearing Stresses on Any Plane Surface

Let the coordinate axes x, y, z be rotated in the directions of three mutually perpendicular axes of principal stress, σ_1, σ_2, σ_3 acting on any plane surface. Let ABC be the plane of unit area which inclines to these coordinate axes. The normal stress σ is supposed to act along the normal to this plane having direction cosines l, m, n and shearing stress τ on it. If S (S_x, S_y, S_z) is the resultant stress on the plane ABC (Figure 9.1) considered at the point P, then

$$S^2 = \sigma^2 + \tau^2, \tag{9.2.1}$$

where $S_x = \sigma_1 l$, $S_y = \sigma_2 m$, $S_z = \sigma_3 n$
so that

$$S^2 = \sigma_1^2 l^2 + \sigma_2^2 m^2 + \sigma_3^2 n^2. \tag{9.2.2}$$

Also, the **measure** of the normal stress σ in the directions l, m, n is given by σ.

$$\begin{aligned} \sigma &= S_x l + S_y m + S_z n \\ \sigma &= \sigma_1 l^2 + \sigma_2 m^2 + \sigma_3 n^2. \end{aligned} \tag{9.2.3}$$

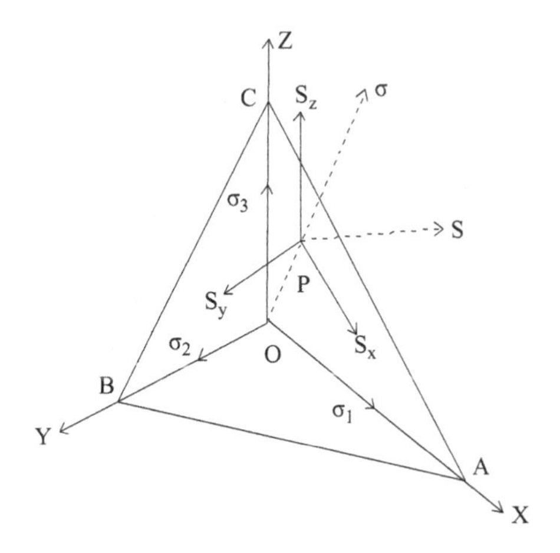

FIGURE 9.1
Normal stress σ and shearing stress τ in any plane ABC of unit area.

Using (9.2.2) and (9.2.3) in (9.2.1), we get

$$\tau^2 = \sigma_1^2 l^2 + \sigma_2^2 m^2 + \sigma_3^2 n^2 - \left(\sigma_1 l^2 + \sigma_2 m^2 + \sigma_3 n^2\right)^2$$

$$= \sigma_1^2 l^2 \left(1 - l^2\right) + \sigma_2^2 m^2 \left(1 - m^2\right) + \sigma_3^2 n^2 \left(1 - n^2\right) - 2\sigma_1\sigma_2 l^2 m^2 - 2\sigma_2\sigma_3 m^2 n^2$$

$$- 2\sigma_1\sigma_3 l^2 n^2$$

$$= \sigma_1^2 l^2 \left(m^2 + n^2\right) + \sigma_2^2 m^2 \left(l^2 + n^2\right) + \sigma_3^2 n^2 \left(m^2 + l^2\right)$$

$$\left(\because l^2 + m^2 + n^2 = 1\right)$$

$$-2\sigma_1\sigma_2 l^2 m^2 - 2\sigma_2\sigma_3 m^2 n^2 - 2\sigma_1\sigma_3 l^2 n^2$$

$$\therefore \tau^2 = l^2 m^2 (\sigma_1 - \sigma_2)^2 + m^2 n^2 (\sigma_2 - \sigma_3)^2 + l^2 n^2 (\sigma_3 - \sigma_1)^2. \tag{9.2.4}$$

This is the required equation for shearing stress τ in terms of principal stress in the direction of the normal stress σ to the plane surface ABC.

To determine the maximum and minimum stresses, we are to study the stationary values of τ from this equation replacing n^2 by $n^2 = 1 - l^2 - m^2$.

$$\therefore \tau^2 = l^2 m^2 (\sigma_1 - \sigma_2)^2 + m^2 \left(1 - l^2 - m^2\right)(\sigma_2 - \sigma_3)^2$$

$$+ l^2 \left(1 - m^2 - l^2\right)(\sigma_3 - \sigma_1)^2.$$

Differentiating it partially with respect to l and m successively, we get

$$2\tau \frac{\partial \tau}{\partial l} = 2lm^2 (\sigma_1 - \sigma_2)^2 - 2lm^2 (\sigma_2 - \sigma_3)^2 + 2l\left(l - l^2 - m^2\right)(\sigma_3 - \sigma_1)^2 - 2l^3 (\sigma_3 - \sigma_1)^2$$

$$= 2lm^2 \left[(\sigma_1 - \sigma_2)^2 - (\sigma_2 - \sigma_3)^2\right] - 2l(\sigma_3 - \sigma_1)^2 \left[1 - l^2 - m^2 - l^2\right]$$

$$= 2lm^2 (\sigma_1 - \sigma_3)(\sigma_1 - 2\sigma_2 + \sigma_3) - 2l(\sigma_3 - \sigma_1)^2 \left(1 - 2l^2 - m^2\right)$$

$$\tau \frac{\partial \tau}{\partial l} = l(\sigma_1 - \sigma_3)[m^2 (\sigma_1 - 2\sigma_2 + \sigma_3) + (\sigma_1 - \sigma_3)\left(1 - 2l^2 - m^2\right), \tag{9.2.5}$$

and similarly,

$$\tau \frac{\partial \tau}{\partial m} = m(\sigma_2 - \sigma_3)\left[l^2 (\sigma_2 - 2\sigma_1 + \sigma_3) + (\sigma_2 - \sigma_3)\left(1 - 2m^2 - l^2\right)\right]. \tag{9.2.6}$$

But for stationary values, $\frac{\partial \tau}{\partial l} = \frac{\partial \tau}{\partial m} = 0$ simultaneously.

Hence, from (9.2.5), $l=0$ for unequal values of σ_1, σ_2, σ_3 (in general) gives when substituted in (9.2.6),

$m=0$ and $m=\pm\frac{1}{\sqrt{2}}$.

Thus,

$$\begin{aligned} &l=0, \quad m=0 &&\text{gives } n=1 \\ &l=0, \quad m=\pm\frac{1}{\sqrt{2}} &&\text{gives } n=\pm\frac{1}{\sqrt{2}} \\ &l=0, \quad m=\pm\frac{1}{\sqrt{2}} &&\text{gives } n=\mp\frac{1}{\sqrt{2}}. \end{aligned} \tag{9.2.7}$$

Similarly, eliminating l and m successively from (9.2.4), we get for other stationary values

$$\begin{aligned} &l=0, && m=1, && n=0 \\ &=\pm\frac{1}{\sqrt{2}}, && m=0, && n=\pm\frac{1}{\sqrt{2}} \\ &l=\mp\frac{1}{\sqrt{2}}, && m=0, && n=\pm\frac{1}{\sqrt{2}}. \end{aligned} \tag{9.2.8}$$

$$\begin{aligned} &l=0, && m=1, && n=0 \\ &=\pm\frac{1}{\sqrt{2}}, && m=\pm\frac{1}{\sqrt{2}}, && n=0 \\ &=\pm\frac{1}{\sqrt{2}}, && m=\mp\frac{1}{\sqrt{2}}, && n=0. \end{aligned} \tag{9.2.9}$$

The first three stationary values of the three sets of (9.2.7)–(9.2.9) are related to principal planes which correspond to minimum values of the shearing stress. The remaining stationary values represent the maximum shearing stresses with respective planes orienting to contain one of the principal axes of stress, which makes 45° angle with the other two axes [7, p. 38] in the plane.

9.3 General Transformation and Maximum and Minimum Longitudinal Strains

Let $P(x, y)$ be any point before deformation which is displaced to the position $Q(x_1, y_1)$ with the general displacement.

$$x_1 = ax + by \text{ and } y_1 = cx + dy.$$

Solving for x and y ($a > 0, d > 0$),

$$x = \frac{dx_1 - by_1}{ad - bc}, \quad y = \frac{ay_1 - cx_1}{ad - bc}. \tag{9.3.1}$$

Let the rectangle joining the points $O(0, 0)$, $M(x, 0)$, $P(x, y)$, and $N(0, y)$ be deformed with the same point P into the parallelogram $ORQS$ with points $(0,0)$, (ax, cx), $(ax + by, cx + dy)$, and (by, dy). It is to be noted that a and d represent the components of longitudinal strains parallel to the x and y coordinates, respectively. Moreover, b and c represent the part shear components characterizing the angular displacements of the initial sides of the rectangle so that

$$ax = l\cos\theta, cx = l\sin\theta,$$

$$\tfrac{c}{a} = \tan\theta$$

$$c = a\tan\theta$$

and

$$dy = l\cos\phi, by = l\sin\phi$$

$$\frac{b}{d} = \tan\phi$$

$$b = d\tan\phi.$$

Now let us investigate what will be the change in a general line $y = mx + p$ after deformation subject to the above general displacement with x, y given by (9.3.1).

∴ Putting the values of x, y in the equation of the line, we get

$$\frac{ay_1 - cx_1}{ad - bc} = m\frac{dx_1 - by_1}{ad - bc} + p$$

$$y_1(a + bm) = (c + dm)x_1 + p(ad - bc)$$

$$\therefore y_1 = \frac{c + dm}{a + bm}x_1 + \frac{p(ad - bc)}{a + bm}, \tag{9.3.2}$$

which is also a straight line.

Hence, the general transformation representing Figure 9.2 characterizes the homogeneous strain.

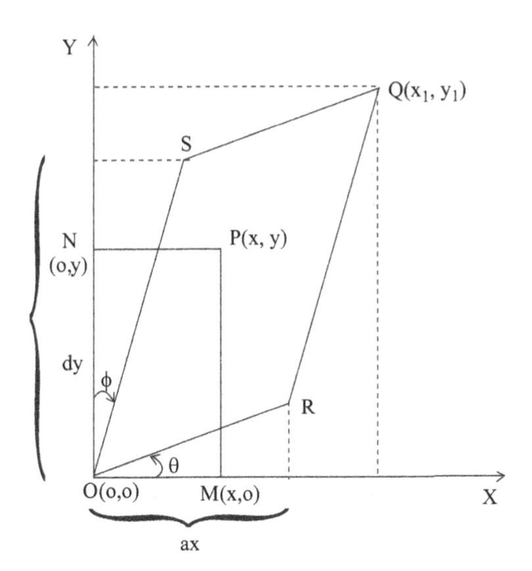

FIGURE 9.2
General strain translation.

Again the points lying on the circle (inscribed rectangle) $x^2 + y^2 = 1$ are subjected to change due to the general transformation, and the circle takes the form:

$$\left(\frac{dx_1 - by_1}{ad - bc}\right)^2 + \left(\frac{ay_1 - cx_1}{ad - bc}\right)^2 = 1$$

$$(c^2 + d^2)x_1^2 - 2(bd + ac)x_1y_1 + (a^2 + b^2)y_1^2 = (ad - bc)^2, \tag{9.3.3}$$

which is an ellipse called **strain ellipse**.

The major and minor axes of this strain ellipse identify the positions of maximal and minimal longitudinal strains. By means of a rotation ψ of axes, the term with x_1y_1 of the equation of strain ellipse (9.3.3) can be removed to reduce it to a standard form, and ψ is given by

$$\psi = \frac{1}{2}\tan^{-1}\left\{\frac{-2(ac + bd)}{(c^2 + d^2) - (a^2 + b^2)}\right\}. \tag{9.3.4}$$

The lengths of the semimajor axis λ_1 and minor axis λ_2 or maximum and minimum strains are given by $1 + e_1 = \sqrt{\lambda_1}$ and $1 + e_2 = \sqrt{\lambda_2}$, respectively, where the strain ellipse is $\frac{x^2}{\lambda_1} + \frac{y^2}{\lambda_2} = 1$.

9.4 Determination of the Two Principal Strains in a Plane

Let $P(x, y)$ be a point with origin O and OP be one of the principal axes of strains so that $OP = 1$. Let θ be the angle made by OP with x-axis. $Q\,(x_1, y_1)$ is the deformed position of P **without rotation**, and hence, for the strain ellipse, the length of one principal strain is $OQ = 1 + e$ (Figure 9.3).

Now from the Figure 9.3, $\cos\theta = x = \dfrac{ax+by}{1+e}$ and $\sin\theta = y = \dfrac{bx+dy}{1+e}$.
($\because$ for irrotational strain $c = b$)

$$\therefore x\left[a-(1+e)\right]+by=0 \tag{9.4.1}$$

and

$$y\left[d-(1+e)\right]+bx=0. \tag{9.4.2}$$

Dividing, $\dfrac{b}{d-(1+e)} = +\dfrac{a-(1+e)}{b}$

$$\therefore \left[a-(1+e)\right]\left[d-(1+e)\right]-b^2=0$$

$$(1+e)^2-(1+e)(a+d)+(ad-b^2)=0, \tag{9.4.3}$$

which gives the two principal **irrotational strains** (the two roots of it), $(1 + e_1)$ and $(1 + e_2)$. From (9.4.1) and (9.4.2), replacing x by $\cos\theta$ and y by $\sin\theta$, we can get

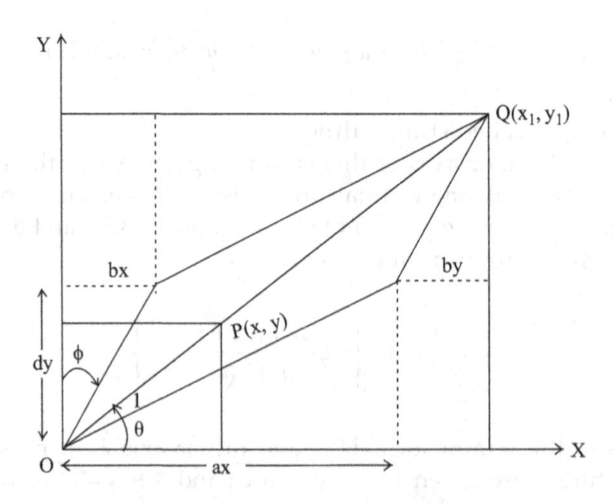

FIGURE 9.3
Graph for principal strains.

$$a-(1+e)+b\tan\theta=0 \text{ and } d-(1+e)+b\cot\theta=0$$

so that

$$\tan^2\theta+\frac{a-d}{b}\tan\theta-1=0, \tag{9.4.4}$$

which is quadratic in $\tan\theta$. This gives the directions of the two mutually perpendicular axes of strain ellipse.

Hence, with reference to the chosen reference axes, the finite displacements (two dimensional) can be instrumental to determine the principal elongation $1+e_1=\sqrt{\lambda_1}$ and $1+e_2=\sqrt{\lambda_2}$ with lengths of principal axes λ_1, λ_2 of the strain ellipse $\frac{x^2}{\lambda_1}+\frac{y^2}{\lambda_2}=1$ derived from the deformation of the initial unit circle $x^2+y^2=1$. To speak the truth, the measures of the two principal strains (magnitudes of strain tensor) can be completely determined from Equations (9.4.3) and (9.4.4), and the angle of rotation ψ by the lines turning to axes of the ellipse.

Finally, to determine the displacements of the points in space resulting in the strain ellipse, significantly no particular coordinate system is chosen in the deformation state. There could have been the same distortions or strains for the displacements with reference to different coordinate systems (Cartesian, of course, in this case) though the equations would have been slightly different. It is discussed in the beginning of the book that the independence of reference frame to study some mathematical concepts is essentially required to use tensors. Quantities such as stresses and strains related to geological context are some second-order tensors or general tensors. For three-dimensional cases, the mathematical entities such as stress will require $(3^2-3)=6$ components (for a symmetric tensor ($\sigma_{xy}=\sigma_{yx}$)) and $3^2=9$ components (for anti-symmetric case). Of course, there are some quantities which remain invariant irrespective of reference frames but need full knowledge of tensor calculus.

The geometrical processes of determination of the measures of stress and strain are discussed here (instead of using truly tensors but magnitudes) to help in its **geological context**.

N.B.: Though actually stress and strain are three dimensional, it is beyond the scope of the book to discuss the general case, and only classical result is reported.

10

Tensors in Fluid Dynamics

10.1 Introduction

Modern scientists and physicists believe that matter is composed of elementary particles, and **in most of the scientific fields**, it is not looked into the individual molecules, which is regarded as an **entity** of small but infinite dimensions interacting with its fellows according to certain laws. So matter is not continuous but discrete, and its gross properties are taken as averages over a large number of molecules. The equations of fluid motion have been formulated from this viewpoint, though they are considered at first sight as much more fundamental one. The average velocity of the molecules is taken in the neighborhood of a point, but how large this neighborhood should be is a questionable one.

In the investigation of ordinary fluid motion, the variations in the medium is considered isotropic (uniform in all directions), and hence, pressure (can be said as classical stress) is a constant quantity. But in case of viscous fluids, it is to be replaced by stress tensor. Eventually, the corresponding body and the surface forces will occur in the system in appropriate forms. The Navier–Stokes equations can suitably meet all the requirements to study the motion of fluids when viscosity is present in the medium.

10.2 Equations of Motion for Newtonian Fluid

Let us consider an arbitrary volume V within the fluid medium enclosed by a surface S.

If ρ is the density of the fluid particles moving with velocity $\bar{v}(v^i)$, then by Reynolds' transport theorem,

$$\frac{\delta}{\delta t}\iiint \rho\, dV = \iiint\left[\frac{\delta\rho}{\delta t} + \rho v^i_{,i}\right] dV$$

$$= \iiint\left[\frac{\partial\rho}{\partial t} + \rho_{,i} v^i + \rho v^i_{,i}\right] dV$$

$$= \iiint\left[\frac{\partial\rho}{\partial t} + (\rho v^i)_{,i}\right] dV.$$

But for conservation of mass within the arbitrary volume dV,

$$\frac{\partial\rho}{\partial t} + (\rho v^i)_{,i} = 0. \tag{10.2.1}$$

This is the **equation of continuity** characterizing the conservation of mass.

If T^{ij} is the contravariant stress tensor on the element ds of the surface with normal n_j, then the force on this element is $T^{ij}n_j ds$.

Also let the contravariant vector f^i represent the external body forces per unit mass of the fluid.

∴ The total net force on the arbitrary volume V enclosed by the surface S when resolved in any direction l_i (say) to it is

$$\iiint_V \rho f^i l_i\, dV + \iint_S T^{ij} n_j l_i\, ds = \iiint_V \rho f^i l_i\, dV + \iiint_V T^{ij}_{,j} l_i\, dV$$

$$= \iiint_V (\rho f^i + T^{ij}_{,j}) l_i\, dV$$

using **Gauss's divergence** theorem to the second term.

But the rate of change of linear momentum is equal to the net external force in the direction l_i.

Hence, $\displaystyle\frac{\delta}{\delta t}\iiint_V \rho v^i l_i\, dV = \iiint_V (\rho f^i + T^{ij}_{,j}) l_i\, dV$

$$= \iiint_V \rho\frac{\delta v^i}{\delta t} l_i\, dV = \iiint_V (\rho f^i + T^{ij}_{,j}) l_i\, dV.$$

But for arbitrary volume V and direction l_i it holds, only when

(for the velocity vector (tensor) $\vec{v}$, $v^i_{,j}$ means covariant derivative)

$$\left(\rho\frac{\partial v^i}{\partial t} = \rho a^i = \right)\rho\left(\frac{\partial v^i}{\partial t} + v^i v^i_{,j}\right) = \rho f^i + T^{ij}_{,j}$$

so that

$$(a^i =)\frac{\partial v^i}{\partial t} + v^i v^i_{,j} = f^i + \frac{T^{ij}_{,j}}{\rho}, \tag{10.2.2}$$

which is the equation for conservation of linear momentum.

Thus, the equations of motion of Newtonian fluids are given by Equations (10.2.1) and (10.2.2).

10.3 Navier–Stokes Equations for the Motion of Viscous Fluids

In the case of ordinary fluids, the stress tensor T^{ij} corresponding to hydrostatic pressure p is $T^{ij} = -pg^{ij}$, and the fluid medium is isotropic. But in the presence of viscosity in the medium, it is anisotropic and is related to deformation.

Let v^i and $v^i + dv^i$ be the velocities of two neighboring points x^j and $x^j + dx^j$.

$$dv^i = \frac{\partial v^i}{\partial x^j} dx^j = v^i_{,j} dx^j.$$

But

$$\begin{aligned} v^k_{,j} = g^{ik} v_{i,j} &= g^{ik}\left[\frac{1}{2}\left(v_{i,j} + v_{j,i}\right) + \frac{1}{2}\left(v_{i,j} - v_{j,i}\right)\right] \\ &= g^{ik}\left[\frac{1}{2}\left(\frac{\partial v_i}{\partial x^j} + \frac{\partial v_j}{\partial x^i}\right) + \frac{1}{2}\left(\frac{\partial v_i}{\partial x^j} - \frac{\partial v_j}{\partial x^i}\right)\right] \\ &= g^{ik}\,(e_{ij} + w_{ij}), \end{aligned} \tag{10.3.1}$$

where

$$e_{ij} = \frac{1}{2}\left(\frac{\partial v_i}{\partial x^j} + \frac{\partial v_j}{\partial x^i}\right) \tag{10.3.2}$$

which is the deformation **symmetric strain tensor** (8.6.4) and

$$w_{ij} = \frac{1}{2}\left(\frac{\partial v_i}{\partial x^j} - \frac{\partial v_j}{\partial x^i}\right) \tag{10.3.3}$$

which is the **rotation tensor** indicating rigid body rotation of the element.

Now the equation of motion with acceleration a^i, stress tensor T^{ij}, and body force f^i is (from Equation (10.2.2))

$$\rho a^i = \rho f^i + T^{ij}_{,j}. \tag{10.3.4}$$

But the stress tensor $T^{ij} = -pg^{ij}$ for viscous fluids is to be supplemented by the viscous stress tensor p^{ij} to result for deformation

$$T^{ij} = -pg^{ij} + p^{ij}.$$

This relationship must be linear and isotropic (same for all coordinate systems) when Newtonian fluid is isotropic.

∴ For the isotropic fourth-order symmetric tensor G^{ijmn} with respect to i, j and m, n, p^{ij} can be written as $p^{ij} = G^{ijmn} e_{mn}$, in terms of the deformation strain tensor e_{mn}.

But for generalization of isotropic fourth-order symmetric tensors, it must be a **linear combination** of $g^{ij}g^{mn}$ and $(g^{im}g^{jn} + g^{in}g^{jm})$ so that

$$\begin{aligned} p^{ij} &= \lambda g^{ij} g^{mn} e_{mn} + \mu(g^{im} g^{jn} + g^{in} g^{jm}) e_{mn} \\ &= \lambda g^{ij} e^m_m + 2\mu e^{ij} \end{aligned}$$

$$T^{ij} = (-p + \lambda e^m_m) g^{ij} + 2\mu e^{ij}. \tag{10.3.5}$$

In the equations of motion (10.3.4), it is necessary to employ stress and strain relation for viscous fluid motion. This requires to express T^{ij} in the following form:

$$\begin{aligned} T^{ij}_{,j} &= [(-p + \lambda e^m_m) g^{ij} + 2\mu e^{ij}]_{,j} \\ &= (-p_{,j} + \lambda e^m_{m,j}) g^{ij} + 2\mu e^{ij}_{,j} \\ &= (-p_{,j} + \lambda e^m_{m,j}) g^{ij} + 2\mu (g^{ik} e^j_k)_{,j} = (-p_{,j} + \lambda e^m_{m,j}) g^{ij} + 2\mu (g^{ik} g^{jm} e_{km})_{,j} \\ &= (-p_{,j} + \lambda e^m_{m,j}) g^{ij} + 2\mu \left[\frac{1}{2} g^{ik} g^{jm} (v_{k,m} + v_{m,k}) \right]_{,j} \\ &= (-p_{,j} + \lambda e^m_{m,j}) g^{ij} + \mu \left(g^{jm} v^i_{,mj} + g^{ik} v^j_{,kj} \right) \quad (j \leftrightarrow k) \\ &= (-p_{,j} + \lambda e^m_{m,j}) g^{ij} + \mu g^{jm} v^i_{,mj} + \mu g^{ij} v^k_{,jk} \\ T^{ij}_{,j} &= (-p_{,j} + \lambda e^m_{m,j} + \mu v^k_{,jk}) g^{ij} + \mu g^{jm} v^i_{,mj}. \end{aligned}$$

But $e^i_m = g^{ki} e_{km} = \frac{1}{2} g^{ki} \left(v_{k,m} + v_{m,k} \right) = \frac{1}{2} \left(v^i_{,m} + g^{ki} v_{m,k} \right)$

$$\therefore e_m^m = \frac{1}{2}(v_{,m}^m + g^{km} v_{m,k}) = \frac{1}{2}(v_{,m}^m + v_{,k}^k) = v_{,m}^m \quad (k \to m).$$

Hence,

$$\begin{aligned} T_{,j}^{ij} &= (-p_{,j} + \lambda v_{,mj}^m + \mu v_{,mj}^m) g^{ij} + \mu g^{jm} v_{,mj}^i \\ &= [-p_{,j} + (\lambda + \mu) v_{,mj}^m] g^{ij} + \mu g^{jm} v_{,mj}^i. \end{aligned}$$

Equation (10.3.4) for viscous fluids takes the form:

$$\begin{aligned} &\rho a^i = \rho f^i - g^{ij} p_{,j} + (\lambda + \mu) v_{,kj}^k g^{ij} + \mu g^{jk} v_{,jk}^i \\ &(v_{,jk} = v_{,kj}), \end{aligned} \tag{10.3.6}$$

which is the **Navier–Stokes equations** for viscous fluids.

Appendix

Some Standard Integrals in Connection with Applications

In many investigations of dynamical scenarios, different kinds of integrals appear to characterize the development of physical situations inviting transformations. The transformation from a volume integral of certain physical property to an integral over bounding surface is a very important frequent measure of classical or tensor analysis. For example, if $\vec{F}$ is the flux of some physical quantity such as flux of fluids or charges, then the integral $\iint \vec{F} \cdot \vec{n}\, ds$ over the whole surface S, where $\vec{n}$ is the outward drawn normal to it, is equal to the total flux out of the closed volume. The Green's theorem (or Gauss's divergence theorem) in connection with such entity can be mathematically stated as follows:

Green's Theorem

Statement: If $\vec{F}$ is any continuously differentiable vector field in volume V and bounded closed surfaces which may have piecewise smooth boundary with outward drawn normal $\bar{n}$, then

$$\iiint_V \nabla \cdot \vec{F}\, dv = \iint_s \vec{F} \cdot \vec{n}\, ds, \quad \text{where } \nabla \text{ is the vector operator.}$$

In tensor form, $\iiint_v F^i_{,i}\, dv = \iint_s F^i n_i\, ds$, for contravariant vector $\vec{F}\left(F^i\right)$.

For covariant formalism,

$$\iiint_v g^{ij} \cdot F_{i,j}\, dv = \iint_s g^{ij} F_j n_i\, ds = \iint_s F_j n^j\, ds.$$

The partial derivatives in Cartesian coordinates are replaced by covariant derivatives, since they are identical.

Stoke's Theorem

Statement: If $\vec{F}$ is any continuous vector field with continuous partial derivatives, then for any two-sided piecewise smooth surface S spanning a closed curve C, $\oint_C \vec{F}\cdot\vec{t}\,ds = \iint_S \operatorname{curl}\bar{F}\cdot\vec{n}\,dS$, where $\vec{t}$ is the tangent vector to the curve C and $\vec{n}$ is the normal right handedly orienting the direction of the curve C.

But in tensor form,

$\iint_S e^{ijk} F_{k,j} n_i\, dS = \oint_C F_k t^k\, ds$, where $\in^{ijk}$ is the permutation tensor.

Reynold's Transport Theorem

If $f(\bar{x},t)$ is any function and $V(t)$ is a closed volume with the fluid consisting of the same fluid particles, then

$$\frac{d}{dt}\left[\iiint_{v(t)} f(\bar{x},t)\,dv\right] = \iiint_{v(t)}\left[\frac{\partial f}{\partial t} + \nabla\cdot(f\,\bar{v})\right]dv \quad \bar{x} = x(x_1,x_2,x_3).$$

Let $\bar{F}(t) = \iiint_{v(t)} f(\bar{x},t)\,dv$, so $\frac{d\bar{F}}{dt}$ is the material derivative which needs to be determined.

$$\therefore \frac{d\bar{F}(t)}{dt} = \frac{d}{dt}\iiint_{v(t)} f(\bar{x},t)\,dv.$$

Since $V(t)$ is variable volume, the differentiation with respect to time cannot be taken under the integral sign. But if we consider the integration with respect to volume in $\bar{\xi}$ – space, $\bar{\xi}$ being material Cartesian coordinate, and change $\bar{\xi}(\xi_1,\xi_2,\xi_3)$ coordinates to $\bar{x}(x_1,x_2,x_3)$, so that

$dV = \dfrac{\partial(x_1,x_2,x_3)}{\partial(\xi_1,\xi_2,\xi_3)}dV_0 = JV_0$ with $dV_0 = d\xi_1.d\xi_2.d\xi_3$ at $t=0$,

then differentiation and integration can be interchanged.

In this case, $V(t)$ can be taken as the moving **material volume** coming from fixed initial volume V_0 at $t=0$ due to the transformation

$\bar{x} = \bar{x}(\bar{\xi},t)$ and $\bar{\xi}$ can be treated as constant.

$$\therefore \frac{d}{dt}\left[\iiint_{v(t)} f(\bar{x},t)\,dv\right] = \frac{d}{dt}\left[\iiint_{v_0} f\{\bar{x}(\bar{\xi},t),t\}J\,dv_0\right]$$

$$= \iiint_{v_0}\left(\frac{df}{dt}J + f\frac{dJ}{dt}\right)dv_0$$

$$= \iiint_{v_0} \left[\frac{df}{dt} + f(\nabla \cdot \overline{v}) \right] J \, dv_0$$

$$\because \frac{dJ}{dt} = \frac{\partial J}{\partial t} + (\overline{v} \cdot \nabla) J$$

$$= \iiint_{v(t)} \left[\frac{df}{dt} + f \, \nabla \cdot \overline{v} \right] dv$$

Associating the gradient terms after making use of material derivatives, it can be arranged to the form:

$$\frac{d}{dt} \left[\iiint_{v(t)} f(\overline{x}, t) dv \right] = \iiint_{v(t)} \left[\frac{\partial f}{\partial t} + \nabla \cdot (f \, \overline{v}) \right] dv,$$

which is the Reynold's transport theorem. The function f can be some scalar or component of tensor.

Corollary 1

Now $\iiint_{v(t)} \nabla \cdot (f \, \overline{v}) dv = \iint_{s} f \, \overline{v} \cdot \hat{n} \, ds,$

(applying Green's theorem)

where S is the surface enclosing the volume $V(t)$.

$\therefore$ The Reynolds' transport theorem can be put to the form:

$$\frac{d}{dt} \iiint_{v(t)} f(\overline{x}, t) dv = \iiint_{v(t)} \frac{\partial f}{\partial t} dv + \iint_{s(t)} f \, \overline{v} \cdot \hat{n} \, ds.$$

Corollary 2

The Jacobian of the transformation from material to spatial coordinates in $dv = J \, dv_0$ is a scalar, and the divergence of the velocity field $v^i_{,i}$ is also scalar.

$\therefore$ (The intrinsic derivative =) $\dfrac{\delta J}{\delta t} = \dfrac{dJ}{dt} = v^i_{,i} J.$

$\therefore$ The Reynolds' transport theorem in tensor form can be written as

$$\frac{\delta}{\delta t} \left[\iiint_{v(t)} f(\overline{x}, t) dv \right] = \iiint_{v(t)} \left[\frac{\delta f}{\delta t} + f \, v^i_{,i} \right] dv$$

$$= \iiint_{v(t)} \left[\frac{\partial f}{\partial t} + f_{,i} \, v^i + f \, v^i_{,i} \right] dv.$$

∵ Intrinsic derivative of scalar invariant $\frac{\delta f}{\delta t} = \frac{\partial f}{\partial t} + f_{,i} v^i$

$$= \iiint_{v(t)} \left[\frac{\partial f}{\partial t} + \left(f v^i \right)_{,i} \right] dv \qquad \text{(i)}$$

$$= \iiint_{v(t)} \frac{\partial f}{\partial t} dv + \iint_{s} f v^i \cdot n_i \, ds,$$

where f is a scalar or component of tensor. Replacing f by ρ (density), the equation of continuity can be recovered from (i).

Remarks

In Chapters 7–10, importance is laid on the theoretical aspects only, showing the use of tensors in the respective branches and that too in restricted senses. Only the fundamental aspects required for initial approach and to pave the way for studying each of the branches which are included in this book are discussed. For this reason, problems are not discussed which would be possible for books of individual subject. The fundamental objective of the author in the book is to report only the glimpses of applications of tensors in some known branches so that readers can in reality enter into the threshold of these branches.

Bibliography

1. P. G. Bergmann, *Introduction to the Theory of Relativity*. Prentice Hall, New Delhi (1962).
2. L. P. Eisenhart, *Riemannian Geometry*. Princeton University Press/Oxford University Press, Princeton, NJ/London (1949).
3. C. E. Weatherburn, *An Introduction to Riemannian Geometry and the Tensor Calculus*. Cambridge University Press, Cambridge (1963).
4. T. J. Willmore, *An Introduction to Differential Geometry*. Oxford University Press, Oxford (1959).
5. R. K. Pathria, *The Theory of Relativity*. Dover Publications, New York (2003).
6. J. B. Hartle, *Gravity: An Introduction to Einstein's General Relativity*. Pearson, New Delhi (2007).
7. J. G. Ramsay, *Folding and Fracturing of Rocks*. McGraw-Hill, New York (1967).
8. G. Ranalli, *Rheology of the Earth*. Allen & Unwin, Boston, MA (1982).
9. M. P. Billings, *Structural Geology*. Prentice Hall, New Delhi (1987).
10. M. R. Spiegel, *Schaum's Outline of Theory and Problems of Vector Analysis and an Introduction to Tensor Analysis*, SI edition. McGraw-Hill, New York (1959).
11. G. E. Mase, *Continuum Mechanics*. Tata McGraw-Hill, New York (2005).
12. R. Aris, *Vectors, Tensors, and the Basic Equations of Fluid Mechanics*. Dover Publications, New York (1989).

Index

M

N

O

P

Q

R

S

T

V

W

Printed in the United States
by Baker & Taylor Publisher Services